THE
SCIENCE - HISTORY
OF THE UNIVERSE

IN TEN VOLUMES

VOLUME 1
FIRST PUBLISHED IN: 1909

FRANCIS ROLT - WHEELER

GIFT *Certificate*

TO:

_____

_____

FROM:

_____

_____

DATE: _____

THE
SCIENCE - HISTORY
OF THE UNIVERSE

VOLUME 1

FIRST PUBLISHED IN: 1909

BY

FRANCIS ROLT - WHEELER
MANAGING EDITIOR

Francis Rolt-Wheeler (December 16, 1876 - August 21, 1960)
was an English writer, astrologer, occultist and esotericist.

DIAMOND
BOOKS
www.diamondbooks.ca

TORONTO, CANADA – 2017

DIAMOND **BOOKS** - CANADA

DIAMOND™
BOOKS

Toronto, ON, CANADA
http://www.**diamondbooks**.ca

BIBLIOGRAPHIC INFORMATION

' THE SCIENCE-HISTORY OF THE UNIVERSE '
was first published in 1909,
by: The Current Literature Publishing Company,
New York.

PUBLISHED IN CANADA

Published in Canada by DIAMOND BOOKS - CANADA, an imprint of
DIAMOND PUBLISHERS - http://www.diamondpublishers.com

REPUBLISHED EDITION: July, 2017.

PAPERBACK EDITION   :   ISBN: 978-1-988942-06-3
E-BOOK EDITION   :   ISBN: 978-1-988942-07-0

PRINTED IN CANADA

THE
SCIENCE - HISTORY
OF THE UNIVERSE

VOLUME 1

FIRST PUBLISHED IN: 1909

ASTRONOMY

By WALDEMAR KAEMPFFERT

--

INTRODUCTION

By PROFESSOR E. E. BARNARD

EDITORIAL BOARD

INTRODUCTIONS BY

PROFESSOR E. E. BARNARD, A.M., Sc.D.,
Yerkes Astronomical Observatory.

PROFESSOR CHARLES BASKERVILLE, Ph.D., F.C.S.
Professor of Chemistry, College of the City of New York.

DIRECTOR WILLIAM T. HORNADAY, Sc.D.,
President of New York Zoölogical Society.

PROFESSOR FREDERICK STARR, S.B., S.M., Ph.D.,
Professor of Anthropology, Chicago University.

PROFESSOR CASSIUS J. KEYSER, B.S., A.M., Ph.D.,
Adrain Professor of Mathematics, Columbia University

EDWARD J. WHEELER, A.M., Litt.D.,
Editor of 'Current Literature.'

PROFESSOR HUGO MÜNSTERBERG, A.B., M.D., Ph.D., LL.D.,
Professor of Psychology, Harvard University.

CONTENTS

vii

The Editor desires to express his gratitude to the universities, the learned societies and the libraries which have placed their facilities at his disposal in connection with this work. Especial thanks are due to the Columbia University libraries, not only for the opportunities afforded, but also for the interest shown in forwarding research work from their collections.

Recognition of courtesy is due to the many publishers who have granted permission for certain quotations from their copyrighted volumes, among them being Messrs. D. Appleton & Co., the Macmillan Co., the S. S. McClure Co. and The McGraw Publishing Co.

To acknowledge the personal indebtedness to the members of the Editorial Board, the Contributors, and all who have assisted with suggestion and advice would make too long a list; but mention should be made of Mr. Edward J. Wheeler, Litt. D., Editor of Current Literature, to whose scholarly judgment and discrimination is largely due what merit may be found herein.   F. R.-W.

INTRODUCTION

In the present volume there have been covered in a comprehensive and popular manner the various departments of Astronomy. Owing to its treatment in a definitely historical and descriptive manner, however, it may be possible to supplement the general review by a few brief statements of some of the results and problems that confront us in the actual work of the observational astronomy of to-day.

There is frequently brought before the astronomer the fact that certain subjects that were apparently exhausted have proved through the more advanced methods of to-day, or perhaps by chance, to be veritable mines of discovery, richer by far than had been anticipated in all the previous investigations. A remarkable illustration of this fact is the splendid work of Professor Hale at the Solar Observatory of the Carnegie Institution at Mount Wilson, California. The Sun had almost been relegated to that limbo from which nothing new can ever come. With the exception of Hale's development of the spectroheliograph, which made possible the continuous photographic study of the surface of the Sun and of the solar prominences, but little advance had been made in solar research for a very long period of time. Even with the new instrument the

ix

work seemed to be confined to the photography of the prominences and a few other features of the Sun that were already observable visually with the spectroscope. Before this the Sun was somewhat of a curiosity and but little new information was had concerning it. It only became really interesting when a total eclipse was imminent, at which time the corona could be seen and studied. The spectroheliograph was the first great step in the study of the Sun. Even though this made possible a continuous photographic record of the prominences and kindred features it could not record the more attenuated and delicate corona. Indeed, we seem to-day as far as ever from any sight of this mysterious object without the aid of the friendly Moon, which for a few minutes at long intervals hides the Sun and gives us our only view of the corona.

But the great work done by Professor Hale and his associates at Mount Wilson (which was foreshadowed by his work at the Yerkes Observatory) in the discovery of the solar vortices and magnetic fields of sun-spots has revolutionized the study of that body and opened up new fields of investigation in this direction that are almost unlimited.

Mr. Abbot, of the Smithsonian Institution, has also established a permanent station at Mount Wilson for the investigation of the solar constant and a general study of the heat of the Sun. The solar investigations, therefore, that are going on at Mount Wilson are among the most important that have ever been undertaken. They are not only of the highest interest, but may ultimately lead to important results bearing upon the commercial life of the world by revealing to us some possible means of forecasting conditions upon the Earth. Any vagaries in the Sun must have more or less direct influence on the conditions of the Earth which owes its every throb of life to the mighty influence of the Sun.

Much of the ordinary spectroscopic work may be said to be in its infancy because of the vast fields of research that are open to it. It is already laying the foundation for a very accurate determination of the distance of the visible binary stars where both stars can be observed with the spectroscope—an accuracy that can never be attained by the ordinary methods of parallax work. Already this has given results of precision in the case of Alpha Centauri, whose distance has been determined by Professor Wright, of the Lick Observatory, from spectroscopic observations combined with the known orbit of the star. Time, however, is an element in this work, and after a sufficiently long interval a valuable harvest of knowledge of star distances will result. The spectroscopic material for such investigations is being specially obtained by Professor Frost and his associates at the Yerkes Observatory (as well as by others elsewhere), where spectrograms of the various visual binaries that are bright enough to give a measurable spectrum are being carefully and accurately accumulated. A possible improvement of the spectroscope, whereby a larger percentage of the light can be utilized, will make possible the extension of this class of work, for at least 90 per cent. of the available light cannot at present be utilized. If this can be done, the efficiency of the spectroscope will be vastly increased and a great number of objects at present beyond the reach of accurate spectroscopic study will be investigated and their nature and physical conditions become known. A step in this direction is the intended erection on Mount Wilson of a reflecting telescope one hundred inches in diameter. The great light-grasping power of this instrument will enable much fainter objects to be studied than can be observed with the present means.

Only a few years ago our knowledge of comets seemed to be satisfactory. What we could see with the naked eye or with the telescope apparently readily agreed with certain theories that were formulated to explain them. The tails of various comets were sorted out and assigned to different classes. This one was a hydrocarbon tail and that a hydrogen tail, etc. The spectroscope had shown that comets in general consisted of some form of hydrocarbon gas (such as cyanogen). Such gas or gases are evidently mixed up with minutely divided matter which is disrupted and expelled from the comet's head and thrown out backward from the comet away from the Sun. This was shown later by the experiments of Lebedew, Nichols and Hull to be due to the pressure of the Sun's light upon the smaller particles of the comet, which drove them away into space with increasing velocity to form the tail. The simple phenomena thus seen by the eye were rather easy of explanation. Photography, however, has revealed such a mass of strange phenomena in these bodies that the theories which seemed so satisfactory before are now seriously questioned, and some of them appear to be entirely inadequate to explain some of the phenomena shown by the photographic plates. But little indication of many of the most extraordinary changes and peculiarities of comets' tails is seen by the eye. In part this is due to the fact that much of the light of a comet is of a nature that has but little effect on the human eye, though it is peculiarly strong in its action on the photographic plate. The first of these bodies to exhibit these peculiarities was Comet IV, 1893 (Brooks). Some of the phenomena of its tail, as revealed on the photographs, appeared to defy the ordinary theories and seemed to show that an influence outside that of the direct action of the Sun upon the comet had manifested itself in the distortion and breaking of the tail. The scarcity of active comets in the succeeding years left this question in abeyance. Comet C, 1903 (Borrelly), however, gave us much information as to the actual velocity of the outgoing particles of the tail, some of which receded from the comet at the rate of 29 miles a second. This object also quite clearly showed that a seat of force of great activity existed in the comet itself, which enabled it to shoot out streams of matter at large angles to the main direction of the tail, which were apparently not bent or affected by the pressure of the Sun's light. The phenomena of Comet IV, 1893, were repeated in Comet C, 1908 (Morehouse). But a great amount of new phenomena was also shown by this last body which demands still greater changes in our ideas of comets and their tails. This object is so recent and its phenomena so startling that astronomers have not yet had time to thoroughly discuss the vast amount of material that exists for its study. Briefly, added to the already known rapid changes in the tail of a comet, this object exhibited the most extraordinary freaks. Tails were repeatedly formed and discarded to drift out bodily in space until they finally melted away. In several cases the tail was twisted or corkscrew shaped, as if it had gone out in a more or less spiral form. Areas of material connected with the tail would become visible at some distance from the head, where apparently no supply had reached it from the nucleus. Several times the matter of the tail was accelerated perpendicularly to its length. At one time the entire tail was thrown forward and violently curved perpendicularly to the radius vector in the general direction of the sweep of the tail through space. This peculiarity is opposed to the laws of gravitation. There is no known cause for this freak of the tail. Evidently we have here, and in many other of the phenomena of this body, some unknown influence at work in the planetary spaces. What this is, is one of the great problems for the future to solve. It has been suggested that many of the unaccountable phenomena of this comet are electrical and can be attributed to the same influence that produces our magnetic storms and auroras on the Earth, and these are believed to be due to abnormal disturbances on the Sun. It is to be hoped that the present return of Halley's comet will add much to a solution of this problem.

The study of the dark or apparently vacant regions of the sky, especially in the Milky Way, is of paramount importance. The photographic plate has shown that the dark regions (the so-called "coal sacks") are generally connected with masses of nebulosity or gaseous matter. These are especially remarkable in the regions of the stars Theta Ophiuchi and Rho Ophiuchi. In the latter case we find a magnificent nebula in a rich region of the Milky Way occupying a hole that is apparently devoid of stars. Some astronomers have attributed the general absence of stars here to absorbing matter—to an opacity and partial dying out of the nebula that cuts off the light of the stars which are beyond it What these apparent vacant regions really are is, therefore, an unsolved problem at present. Some of them are evidently due to the thinning out and actual absence of stars in those parts of the sky. But the others, which are connected with nebulosities, seemingly must have some other explanation. One fact that appears to be brought out by the great nebula of Rho Ophiuchi is that the groundwork of the Milky Way in this region, and by inference elsewhere, may be made up of stars actually much smaller than the average of those seen in the general sky. If this were so it would materially change our ideas of the Milky Way. This supposition comes from the fact that the great nebula is connected with some of the brighter stars in this region, while at the same time there is apparently evidence that it is connected with the faint stars that form the groundwork of the Milky Way here. If, however, the dark regions about and near the nebula are due to the absorption of light by an opacity of the nebula, the supposition as to the relative sizes would not hold, for the nebula in that case might be very much nearer to us than the Milky Way. It will be evident that an understanding of the nature of the dark regions of the Milky Way is of the utmost importance to a proper knowledge of our stellar universe.

The great nebulous regions of the sky that photography has revealed to us are intimately connected with the Milky Way. They cover very large regions of the heavens and must be almost inconceivably great. In no case has it been possible to determine the exact dimensions of these wonderful objects, because we do not know their distances. It is possible, however, by assumptions that are justified by facts to arrive at some idea of their minimum extent. If they are no further away than the nearest fixed stars, and from their evident connection with certain stars we know that they must be much further away, we can form some idea of their vastness. Our own Sun if removed to the distance of the nearest fixed stars would present an apparent diameter of about the hundredth part of a second of arc. Its known diameter is something like a million miles (accurately 867,000 miles). Some of these nebulous regions are many degrees in diameter. The one connected with the Pleiades is ten degrees in diameter. It is certainly connected with the cluster whose distance is much beyond the nearest fixed stars. From this it will be readily seen that this great nebulous region must be at least some four million times greater in diameter than our Sun, or over one hundred thousand times greater than the entire diameter of our known solar system. These are figures that appear to be appallingly great. But they are only relatively so and only shock us because the facts are new and we are not yet used to them.

What is the ultimate function of these enormous masses of gaseous matter that we find lying in space? Are we sure that they are the primitive matter from which worlds and systems are finally to be evolved?

These, very briefly, are a few of the problems that we encounter in astronomy as developed by the subtle means of research in use at the present time.

E. E. BARNARD.

ASTRONOMY

CHAPTER I

THE EVOLUTION OF ASTRONOMICAL IDEAS

HERBERT SPENCER has stated that evolution is a change from the indefinite to the definite, from the incoherent to the coherent. If any proof of that doctrine were required, it would assuredly be found in the development of astronomical conceptions. In this chapter an attempt will be made to outline in a general way the manner in which the present theories were evolved from the mysticism of folk-lore and religion. Some of the matter herein presented is drawn from Arrhenius' "Die Vorstellung vom Weltgebäude im Wandel der Zeiten."

The astronomical beliefs of prehistoric man were no doubt similar to those entertained by the Eskimo of the Arctic regions and the untutored tribes of Argentine Republic, South Africa and Australia, tribes who, living only for the day, concern themselves but little with to-morrow and yesterday and care nothing about the universe.

Somewhat more cultured than these Eskimo and South American and South African tribes are primitive nations who have endeavored to account for the origin of the Earth and the heavens by anthropomorphic theories. The universe must have been created by some Personal Being who had at his disposal something to mold. The idea that the universe was made out of nothing is a philosophical assumption which was introduced by the highly cultured philosophers of the East. The something out of which the universe was created is usually regarded as water, an element which to the eye at least is perfectly homogeneous, shapeless, and chaotic. That the fertilizing mud was deposited by floods must have attracted the attention of ancient primitive races, for which reason they may have assumed that all the Earth was slowly and gradually deposited from water. Thus we find that Thales (550 B.C.) argued that all things were created from water. Yet other substances were assumed as primordial matter, and later Anaximines of Miletus, who also flourished in the sixth century, called the generative principle of things air or breath, while Heraclitus, who flourished at Ephesus near the end of the sixth century, believed that all bodies were transformations of one and the same element, which he called fire.

The belief that primordial water is the origin of all things was deeply rooted in Asiatic races, for it occurs over and over again in many creation myths, among others in the Chaldean and in the Hebrew. Instead of water we sometimes find that an egg may be taken as the primal unit, no doubt because every organism springs from an apparently lifeless seed. Thus we find that the egg plays a most important part in the creation myths of the Japanese as well as in narratives from India, China, Polynesia, Finland, Egypt and Phenicia.

In many of these creation myths, of which I. Riem has collected no fewer than sixty-eight, more or less independent of one another, deluges are prominent features. In nearly all of them it is supposed that after the water subsided the land was exposed, fertilized and made to bring forth.

All of these creation myths are interwoven and interconnected with religious belief. To the savage mind everything that moves is endowed with a Spirit. Accordingly primitive man endeavors to propitiate the Spirit by magic, knowledge of which art is given only to the medicine man or to the priest. In a certain sense, therefore, magic is the precursor of natural science, and the myths and lore upon which the practice of magic is based are remotely antecedent to our scientific theories. According to Andrew Lang, myths are based as much upon primitive science, resting upon superstition, as upon primitive religious conceptions. •

In Maspero's "Histoire Ancienne des Peuples de l'Orient Classique" we find an account of the Chaldean conception of the universe. Surrounded on all sides by the ocean, the Earth rises in the middle like a high mountain whose summit is covered with snow from which the Euphrates springs. The Earth is encircled by a high wall, and the abyss between the Earth and the wall is filled by the ocean. Beyond it is the abode of the immortals. The wall supports the vault of the firmament, shaped by Marduk, the Sun god, out of a hard metal, which shines in the daytime but which at night is like a blue bell set with stars. In the morning the Sun enters the vault by an eastern entrance and at night makes its exit by a western outlet. Marduk arranged the year according to the course of the Sun and divided it into twelve months, each of which counted three periods of ten days. The year, therefore, numbered three hundred and sixty days. Every sixth year a special year was intercalated, so that the year had on an average three hundred and sixty-five days.

As the lives of the Chaldeans were to a high degree influenced by a change in the seasons, they laid great stress upon division of time. In the beginning they probably based their chronology upon the movements of the Moon, like many another race. Soon they recognised that the Sun exerted a stronger influence, and accordingly they introduced a solar year whose divisions they ascribed to Marduk. The stars were observed because their positions determined the seasons. Since the seasons govern organic life, a pernicious belief in the influence of the stars took root, a belief which prevailed for twenty centuries and which crippled the advance of science up to the time of Galileo. Diodorus Siculus, a contemporary of Julius Cæsar, describes this astrology in the following words, as given in a translation by Philemon Holland (1700):

"Therefore from a long observation of the Stars, and an exact Knowledge of the motions and influences of every one of them, wherein they excel all others, they (the Chaldean astrologers) foretell many things that are to come to pass.

"They say that the Five Stars which some call Planets, but they Interpreters, are most worthy of Consideration, both for their motions and their remarkable influences, especially that which the Grecians call Saturn. The brightest of them all, and which often portends many and great Events, they call Sol, the other Four they name Mars, Venus, Mercury, and Jupiter, with our own Country Astrologers. They give the name of Interpreters to these Stars, because these only by a peculiar Motion do portend things to come, and instead of Jupiters, do declare to Men beforehand the good-will of the gods; whereas the other Stars (not being of the number of the Planets) have a constant ordinary motion. Future Events (they say) are pointed at sometimes by their Rising, and sometimes by their Setting, and at other times by their Colour, as may be experienced by those that will diligently observe it; sometimes foreshewing Hurricanes, at other times Tempestuous Rains, and then again exceeding Droughts. By these, they say, are often portended the appearance of Comets, Eclipses of the Sun and Moon, Earthquakes and all other the various Changes and remarkable effects in the Air, boding good and bad, not only to Nations in general, but to Kings and Private Persons in particular. Under the course of these Planets, they say are Thirty Stars, which they call Counselling Gods, half of whom observe what is done under the Earth, and the other half take notice of the actions of Men upon the Earth, and what is transacted in the Heavens. Once every Ten Days space (they say) one of the highest Order of these Stars descends to them that are of the lowest, like a Messenger sent from them above and then again another ascends from those below to them above, and that this is their constant natural motion to continue forever. The chief of these Gods, they say, are Twelve in number, to each of which they attribute a Month, and one Sign of the Twelve in the Zodiack. Through these Twelve Signs the Sun, Moon, and the other Five Planets run their Course."

The Chaldean priests developed a most perfect astrology. They mapped out the positions of the stars for every day with such care that they could tell their true positions for some time in advance. The different stars either represented deities or were directly identified with them. If, therefore, a Chaldean king wished to know which gods ruled over his destiny, he consulted the priests as to the position of the stars on his birthday and was informed of the chief events of his career.

This Chaldean belief that the celestial bodies were gods transformed astronomy into a religion. Hence astronomical theories were promulgated only by the ruling priest caste. To doubt the tenets of that caste was to expose oneself to merciless persecution, an Oriental trait that passed over to the nations of classic antiquity and to the semi-barbarous nations of the Middle Ages.

The Jews appropriated the Chaldean conception of the universe, but modified it, so that it was transformed from a polytheistic to a monotheistic conception.

No doubt the Chaldaic accounts of the beginning of the world influenced Egyptian thought. According to Maspero, the Egyptians believed that matter without form was shaped by a deity, always a different person in different parts of the land and by different methods, into the world as we see it.

The classic nations borrowed much of Egyptian civilization and with it Egyptian religion and science. For, the Greek creation myth, like all the others, assumes that chaos once existed and that out of it Gaa, the mother of all things, and her son, Uranos, the god of heaven, were created.

The Greek cosmogony was adopted by the Romans without noteworthy development. Hence it is that Ovid wrote on the origin of the universe much as Hesiod had done seven hundred years before. In that long interval of seven centuries the study of nature had advanced but little. Indeed it was not until the invention of the telescope that astronomy was lifted entirely out of the hands of the priesthood and placed upon a sure scientific footing. Before the invention of the telescope, therefore, astronomy appears merely in the garb of a myth. At its best it was metaphysical.

The rudiments of astronomical science are to be found in the efforts of the Chaldeans, Egyptians and Greeks to devise calendars and to mark time. That effort necessitated a study of the motions of the celestial bodies. Moreover, exigencies of husbandry rendered necessary some method of keeping track of the seasons so that seed time and harvest could be ascertained. The regular occurrence of such events as the Nile flood made requisite suitable preparations. Hence the early Egyptians so built their temples that they might know the time of the summer solstice and hence the time when the flood might be expected. This was a matter of practical importance, not merely connected with religion or priestcraft, but on which the lives and the happiness of the people of Egypt depended, and might be compared with the modern time observations made at the great national observatories. The observation of the stars was carried on with at least this object in view, and gradually with the development of civilization time reckoning from the stars became an important consideration closely connected with the lives of the people. With the study of the stars for such a purpose naturally an amount of information as to their positions and motions was accumulated, and for centuries the practical side of astronomy was the study of the position of the stars and the motion of the planets. The astrology of the Chaldeans spreading westward increased rather than diminished the interest in the stars, for not only was the connection of the planets with natural phenomena and the mere reckoning of time studied, but the mystical element involving prophecy of future events attracted attention. In other words, astrology was a pseudo-science, for which reason it is difficult to estimate its benefits or to exaggerate its evils. In its scientific aspect it involved the observation and record of the position of the heavenly bodies with all the exactness that the mathematical and observational methods of the time could achieve. It enabled the motions of the planets to be studied as well as the positions of the fixed stars and the course of the Sun as it passed through them. But, on the other hand, when the interpretation of the appearance of the skies was involved, superstition and poetic fancy had full sway, in which no doubt certain elements of self-interest and deception on the part of the priests or astrologers were not lacking. Hence these men did not study the sky to interpret phenomena on a scientific basis. Confined in the narrow limits of superstition, they not only made no progress but actually held back astronomy as they did other sciences.

That the work of the astrologer was mysterious there can be no doubt, and as no reason was assigned for the movement of the planets or the position of the stars, it was a natural assumption on the part of the people that some supernatural agency was at work, which also was connected with their lives and their future. With the beginning of the development of scientific astronomical theory proper the power and position of the astrologers began to wane—slowly, it is true, for when Tycho Brahe was invited to deliver lectures on astronomy at the University of Copenhagen, the first dealt very largely with astrology. Cardan and Kepler among the distinguished astronomers of the Middle Ages, Roger Bacon, Burton and Sir

Thomas Brown were among the men of mind who were interested, at least in part, in the teachings of the underlying basis of the cult. As explanations of the motions of the heavenly bodies on a rational basis were forthcoming, the doom of the astrologer, so far as participation in the scientific creed of the day was concerned, was sealed. If there was a natural explanation that could be accepted, how could supernatural influences condition the movements of the planets or the positions of the stars? If then these movements were natural and made in obedience to natural laws, how could they affect the future course of life and future occurrences that obviously had no connection with natural phenomena? The law of gravitation, which explained the solar system and the movement of the planets, corroborated this view and left only the comets as striking natural phenomena which could not be explained in a way that the popular mind could grasp. With the rise of learning and the growth of observation, the explanations of natural phenomena by astronomers secured acceptance by the people. Finally, when Halley's prediction of the return of his comet, first made in 1705, was verified in 1758, the reign of natural law in the world of the heavens became an accepted fact, from which only the ignorant or superstitious could dissent.

Distinctly different and apart from astrological influence was the work of Copernicus, whose researches mark the beginning of the new and philosophical science of astronomy, in which the element of mysticism was gradually displaced and observation and reasoning were depended on. Copernicus, as will be seen when the development of theories of the solar system is considered in an early chapter, returned to many of the fundamental ideas of Pythagoras and the early Greek philosophers, especially that the Sun was the center of the universe. He was a thoughtful student not only of Greek philosophy but of the work of such later astronomers as Ptolemy and his successors, so that when he announced a theory of the solar system in which the Earth and other planets revolved around the Sun as a center, it was based upon the fullest knowledge of previous reasoning and theory. Nevertheless he was casting to one side the tradition and the science of the day as it was then understood and presenting what was a conception of the heavenly world no less daring than original. His theory was a natural outcome of the revival of learning in the Renaissance, foreshadowed by the work of such men as Leonardo da Vinci and others, in whom the scientific spark had been awakened. With Copernicus the evolution of his heliocentric theory was a matter of scientific reasoning rather than of direct observation. But it marked the beginning of a series of epoch-making discoveries presented in a clear and positive form, so that the theory of the revolution of the planets around the Sun became one of the fundamental canons of astronomy. Thus, as will appear in the course of our history, the Copernican theory in which the revolution of the planets around the Sun is made clear, Kepler's theory of planetary motion in which laws are stated to account for this motion, and finally, Newton's announcement of the great universal law of gravitation, are the foundation stones on which modern astronomical science firmly rests.

The invention of the telescope established the similarity in the bodies of the solar system and revealed facts that previously had been hidden from observers of the heavens. Indeed, with the invention of the telescope and the growth of mathematical science, there began an era of descriptional astronomy in which exact observation was combined with careful computation and mathematical analysis, an era which continued into the nineteenth century with undiminished vigor. Brilliant discoveries were made possible by improved and powerful instruments, accompanied by theoretical work of even greater value. In the middle of the nineteenth century new instruments were put at the command of the scientist which had a remarkable effect in extending the boundaries of the science. The telescope had facilitated merely the observation of the stars. The spectroscope, on the other hand, enabled the astronomer to ascertain their composition.

With the application of the spectroscope to astronomy began the welding of physics and chemistry with astronomy and the birth of that modern science of astrophysics, which has afforded data for the study of the serious problems connected with the evolution of the universe. From the soothsaying star-gazer of Chaldean times to the modern astrophysicist, who works in a laboratory as well as in an observatory, we have a development that is responsible for the aggregation of knowledge which we now possess of the vast universe with its suns, planets, stars and nebulæ.

The spectra of distant celestial bodies recorded on the photographic plate by the spectrographs of large telescopes are now studied in comparison with the spectra of terrestrial substances produced in the physical laboratory. Not only the nature and composition of the stars can be ascertained, but also their motion in space which are beyond the range of any telescope. The New Astronomy has become on its astrophysical side almost an experimental science with the methods and accuracy of the chemical or physical laboratory. It is from this modern astronomy, with its breadth and resourcefulness, that modern science looks not only for advances in its own particular field, but in the broader and ever interesting problems of cosmogony as concerned in the evolution of the stars and other bodies making up the universe.

CHAPTER II

THE history of astronomical observation is the history of man's attempt to bring the stars nearer to him. His own senses are so feeble and so very subject to error that he has been constrained to devise subtle artificial senses which take the place of eyes and hands. Thus early he invented position-finders, which enabled him to determine with more or less precision a star's direction or position at a given time and not merely to guess at that position; great eyes, called telescopes, that see what his eyes can never see and also determine positions with greater accuracy; wonderful spectroscopes that analyze a star's composition as nicely as if it were a stone picked up in the road; and photographic devices that reveal secrets of star structure that otherwise would never be disclosed by his unaided senses.

For determining the position of the heavenly bodies the instruments used have always been comparatively simple. All are based on certain rudimentary geometric principles. As geometry was a science fairly well developed among the ancients, it is not difficult to realize that they had various means of measuring angles, both vertical and horizontal. In most ancient cases, however, the observers have failed to hand down their methods, merely recording the results without indicating the circumstances in which they were obtained, so that it is impossible to discuss the values of the observations and correct them in the light of recent discoveries. It is evident that the instruments of the ancients were simple, but their precise nature is altogether uncertain.

The earliest astronomical observations of which there is record were made by the Chinese. The Shu King, the oldest known scientific work, states that two thousand years before the present era the Chinese determined the seasons—that is to say, the positions of the Sun at the equinoxes and solstices—by means of four stars which have since been identified and found to be so suitable that a modern astronomer could not have made a better choice. The Chinese also determined, eleven hundred years before the present era, the obliquity of the ecliptic, which they found equal to 23 deg. 54 min. The obliquity, which varies, is now 23 deg. 37 min., and calculation shows that at the epoch of the Chinese observations it must have been 23 deg. 51 min. Hence the error of the Chinese determination was only three minutes of arc.

Among the few astronomical values which have remained constant during the history of man are the times of revolution of the planets. The Hindus determined the revolution of Mercury with an error of $\frac{1}{10000}$ of a day. For Venus the error was $\frac{28}{10000}$ of a day, for Mars $\frac{1}{1000}$ of a day. In the case of Jupiter the error amounts to one-quarter of a day, but it is to be remembered that the period of revolution of this planet exceeds 11 years, so that the same observer could not observe many returns of the planet to the same point of its orbit. This comment applies with still greater force to Saturn, the revolution of which occupies 29 years. Hence it is not astonishing that in this case the Hindus were six days in error.

Among the ancient Greeks is a measurement of a terrestrial meridian made about 200 B.C. by Eratosthenes (276 B.C. to 195 or 196 B.C.), who found the circumference of the Earth equal to 250,000 stadia by measuring the angular distance of the Sun from the zenith at the summer solstice both at Alexandria and at Syene in Upper Egypt by means of the length of shadow cast by a vertical pillar at noon at each place. According to the researches of Tannery, the stadium as an astronomical unit equals 157.5 meters (516.7 feet), which gives for the Earth's circumference a length of 39,690 kilometers (24,662 miles) instead of 40,000 kilometers (24,855 miles) as we know it. Here the precision is remarkable, especially when it is remembered that the measurement was effected by counting the paces contained in an arc of the meridian and by multiplying the number so found by the length of a pace.

The instruments most frequently employed by early astronomers were divided circles and compasses with simple sights which allowed the line of vision to be directed to the star under observation and its direction as compared with some other line of sight to be measured. Ptolemy's ring or astrolabe, for example, described in the fifth book of his Almagest, and used to identify the relative positions of the stars and planets, was composed of two concentric vertical circles. The outer circle, about 16 inches in diameter, was fixed and graduated. It supported the interior ring, which was movable and carried the two sights. There was also a geometric square which was used in a manner analogous to that of a table of logarithms. Various forms of apparatus for the measurement of horizontal and vertical angles were early evolved, and as the study of the heavenly bodies developed to a point where it was useful in navigation, the cross-staff or back-staff was invented, consisting of simple sighting bars with crosspieces suitable for the calculation and measurement of such angles as the heights of the heavenly bodies above the horizon and their distance from one another. Quadrants of one form or another, with a sighting bar and divided circular scale, and astrolabes, or celestial circles, also for the direct measurement of angles, were employed. Many of these, by the Middle Ages, were examples of accuracy of division.

A quadrant designed by Tycho Brahe (1546-1601), for

example, was of 19 feet radius and had its circumference graduated to single minutes. Various forms of armillary spheres were constructed in which the stars were placed in their relative positions on great circles of the celestial sphere. Such devices served for much of the early astronomical work, taking the place of modern star charts.

Tycho Brahe, like his predecessors, employed wooden instruments. One of these was a large Ptolemy's ring, surmounted by a post carrying horizontal arms, by which it was turned in bearings like a capstan, so that the ring could be brought into any vertical plane. Tycho Brahe also constructed a mural circle, by means of which vertical angles could be measured. Hence it was by using the naked eye and rudimentary instruments that he accumulated observations of such precision that they served Kepler as the basis of the researches which led to the discovery of the laws of planetary movement.

The eye can distinguish an object whose diameter is equal to about $\frac{1}{3000}$ of its distance, which corresponds with an angular diameter of about one minute of arc. This was the measure of the precision of early observations. Its value may be appreciated by stating that it corresponds with the diameter of a lead-pencil seen at a distance of 70 feet. The telescope, by increasing the distance at which objects can be distinguished, therefore has been and is now the chief reliance of the astronomer in determining position. While the naked eye to-day may be said to have been very largely supplanted by spectroscopic and photographic observation, yet the telescope has constantly met the demands of astronomers as its power has increased and its scope widened.

By chance or otherwise it was found by a Dutch spectacle maker, Lippershey, about 1608, that two lenses when placed at some distance apart would act to magnify distant objects, just as a single lens would enlarge the image of a near-by object. This action of the lens can be explained by considering the effect on a prism of transparent material placed in the path of a beam of light. When a beam of light falls on one of the angular faces of the prism at a direction other than perpendicular to the face it is forced to change its direction on account of refraction, due to the change in medium. That is, a ray of light passing obliquely through air into a denser medium, such as glass, is bent toward the perpendicular and in passing out from a denser to a rarer medium is bent away from the perpendicular. A lens may be considered as a collection of prisms of constantly changing angles, so that the effect would be to bend parallel rays coming from a point at infinite distance in such a way that they would all be brought to a single point known as the focus. Consequently a telescope may be regarded as a light-gatherer.

The importance to astronomy of Lippershey's invention can be appreciated from the fact that as soon as Galileo heard of it he constructed such an instrument which, hardly the size of a small toy spyglass, magnified three times, or brought the heavenly bodies three times as near. He applied it to celestial observation in 1609.

The value of the telescope as an astronomical instrument became apparent immediately. It was from the use of his "optik tube," as he called it, that Galileo arrived at the conclusion that Ptolemy was wrong and Copernicus right —how will become apparent from a consideration of the discoveries made by Galileo. He did more than this, however; for by the application of the telescope to the observation of the stars he became in truth the founder of our modern science of astrophysics.

Galileo saw hosts of stars never before revealed to the unaided eye. The six stars in the Pleiades now appeared as 36, and various nebulous objects of light, such as the Milky Way, were found to consist of multitudes of fine stars clustered together. But his crowning achievement occurred on January 7, 1610, when in turning his telescope toward Jupiter he discovered four satellites of that planet and determined that their periods of revolution around

Jupiter ranged from about forty-two hours to seventeen days. Here was a miniature system similar to that conceived by Copernicus. Was it any wonder that Galileo abandoned the Ptolemaic teaching? Thus Galileo was able to strike a serious blow at the infallibility of Aristotle and Ptolemy, by whom no mention had been made of the existence of such extra bodies. At this time, however, others besides Galileo were working with the telescope, among them Thomas Harriott (1560-1621) in England, Simon Marius (1570-1624) and Christopher Scheiner (1575-1650) in Germany. Thenceforth observational astronomy with the telescope was anchored on a firm basis.

As was quite natural, telescopes eventually formed an important part of the equipment of the observatory of Tycho Brahe and of John Kepler (1571-1630). In one of Kepler's works on "Optics" is contained a suggestion for the use of a convex lens for an eye-piece in the construction of the telescope. Galileo's instrument consisted of a lead tube containing a large double-convex lens, which served as an objective, and a small double-concave lens at the eye-end in order to give an erect image—an arrangement which finds its counterpart in the modern opera glass.

Kepler's suggested improvement provided a more efficient and fairly modern astronomical telescope. The actual construction of an instrument of this type, however, is credited to Scheiner rather than Kepler, who was notably deficient in mechanical skill. After considerable experimenting by various astronomers and instrument makers, it was found that a comparatively small objective with a considerable focal length was most useful and effective. In 1672 Capani, of Bologna, constructed an instrument of this kind 136 feet long, while Auzout actually made a telescope 600 feet in length, which, however, failed to work. These, of course, were skeleton structures not mounted in tubes. Perhaps the best of them were those of Huygens

(1629-1695), whose skill in grinding lenses stood him in
such good stead that he was able to construct a telescope
with which he determined the ring of Saturn. Huygens'
telescope had considerable focal length. He placed the
object glass on a tall vertical pole or staff so balanced that

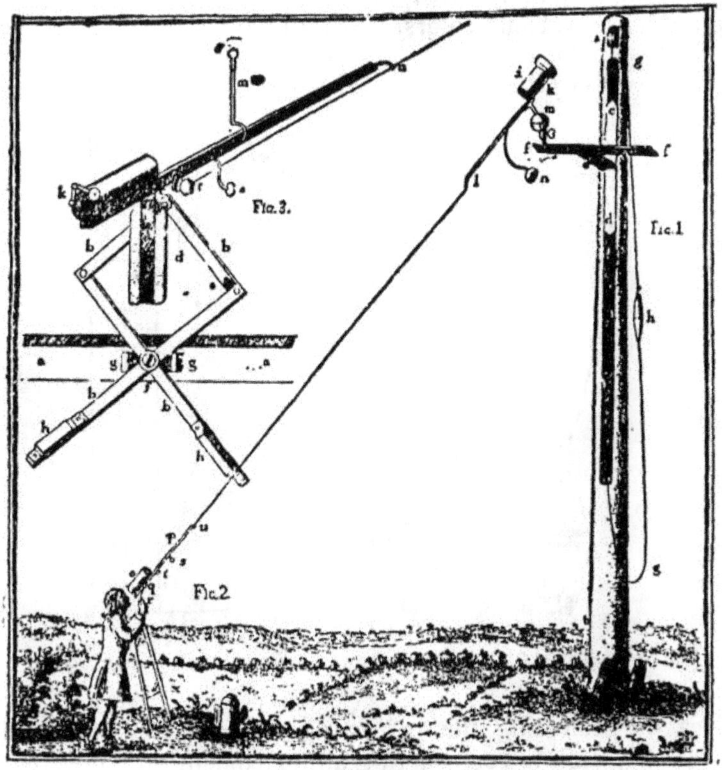

Fig. 1 —Huygens' Aerial Telescope.

it could be moved in any direction by means of a cord.
The observer on the surface of the Earth was supplied
with an eye-piece which he maintained in a straight line
with the star he was observing by means of a cord.

All these telescopes were "refractors." They were subject to certain inherent defects, chief among which was the difficulty of bringing to a single focus all the rays of different colors. The seventeenth-century philosophers believed it impossible to overcome the unequal refrangibility of the different colored rays of light which produced "chromatic aberration" and resulted in an image indistinct for the blurring of various colors. Accordingly they gave up the

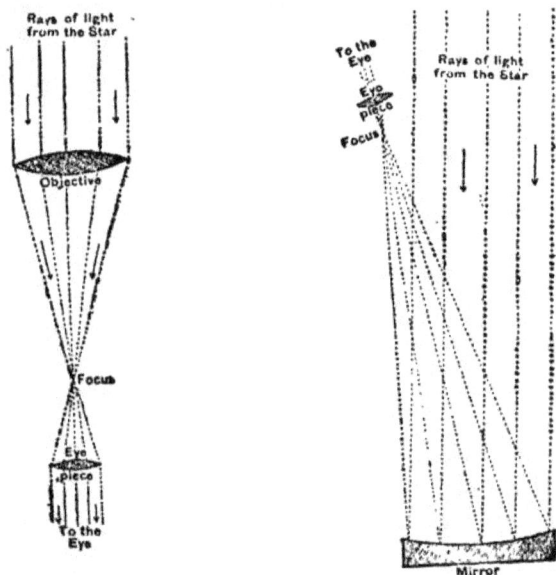

Fig. 2 —REFRACTING TELESCOPE.   REFLECTING TELESCOPE.

idea of perfecting the refracting telescope and directed their attention to constructing an instrument on a different principle, using a concave mirror to form the image of the object observed. Mersenne, in 1639, suggested the employment of a spherical mirror, but the idea appears to have been dropped. Quite independently, James Gregory, in 1663, proposed a similar arrangement, using, however, a parabolic in place of a spherical mirror. At that time he could not find a workman able to construct such a mirror.

In the Gregorian instruments the parabolic reflector is placed at the lower end of the tube, while on its axis and a short distance beyond its focus is placed a small concave reflector. The light from the distant object falls upon the large mirror, from which it is reflected back to the small one, which throws it back through a hole in the center of the large reflector. It then passes into the eyepiece, which, indeed, in Gregory's time had been much improved by Huygens.

Gregory's efforts turned Newton's attention to reflecting telescopes. In 1669 he cast his first disk and began to grind it, but it was not until 1672 that he had real success. Then he made two small instruments, one of which was only about an inch in diameter, with a magnifying power of about 38. The principle of Newton's telescope differed from that of Gregory's in that it had a small plane mirror placed in the cone of light from the reflector at an angle of 45 degrees. Being placed inside the focus, this mirror brought the light cone at right angles to its original direction, thus forming the image outside the tube and obviating the necessity of a hole in the parabolic reflector.

About the same time that Newton completed his instrument the Cassegrain construction was proposed, which was similar in construction to the Gregorian telescope, except for the small mirror. In the Gregorian this mirror was concave. Cassegrain (1672) proposed the use of a concave mirror inside the focus. It brought the light from the object to a focus through a hole in the center of the large parabolic mirror.

Owing to the difficulty in obtaining a suitable mirror alloy, little progress was made for some time in constructing reflecting telescopes. In 1718, however, Hadley, the inventor of the sextant, constructed one on the Newtonian principle, 5 feet in length. The instrument magnified over 200 times and revealed as much as the old refracting telescopes. Perfect as this Newtonian telescope seemed to be, the Gregorian type held the field until 1774.

By using a small Gregorian telescope Herschel had his attention directed to the wonders of astronomy. His income being too limited to purchase an instrument, he set about making one for himself. During his life he is said to have made upward of 400 telescopes, mostly of the Newtonian type. Among his earliest efforts was the construction of a 5-foot reflector, which was a wonderful success. Then came one 7 feet in length. The largest of his instruments was completed under George III. in 1789. This telescope surpassed all his previous efforts, for it was actually 40 feet long and had a reflecting mirror 4 feet in diameter. The story of Herschel's work with this great telescope would fill a volume.

The largest telescope of the reflecting type was constructed by Lord Rosse, an Irish peer, and used at Parsonstown. It had mirror of 54 feet focus and a diameter of 6 feet, but it could be used only for observations on or near the meridian. While out of use for many years, it long held the record for size, which, however, is now taken by the 100-inch reflector recently completed for the Mount Wilson Solar Observatory, California.

The case of the refracting telescope, which as we have seen had been all but abandoned on account of its chromatic aberration by the seventeenth-century astronomers and physicists, was not as hopeless as they believed. Notwithstanding Newton's dictum that it was useless to try to improve it, owing to the impossibility of producing refraction without dispersion, Euler read a paper before the Berlin Academy in 1747 proving mathematically the possibility of correcting both the spherical and chromatic aberration of an object glass. Upon reading Klingenstierna's paper corroborating Euler's views, John Dollond made a series of most valuable experiments which led him to the solution of the problem of the achromatic object glass—namely, that by properly combining two kinds of glass, flint and crown, he could unite the colored rays fairly well and still have refraction to unite the incident rays to form an image.

Dollond's discovery occurred in 1758; his work soon became famous. He was surely master of his subject and had a clear field for many years. Like other opticians, he labored under great difficulties in securing glass suitable for telescopes of any diameter. Fortunately a genius had taken hold of this problem in the person of Guinand, a Swiss watchmaker, who, after long experimenting, solved the problem of making fine disks of optical glass. He associated himself with the celebrated Fraunhofer in 1805, and they successfully made optical glass disks up to fifteen inches aperture. To Fraunhofer are due many of the most important discoveries in the theory of the achromatic objective.

With proper optical glass and methods of correction the refracting telescope soon came into its own. The size of the objectives was increased so that sufficient amounts of light were gathered to form a distinct image. The best makers of Europe gradually developed both lenses and mountings so that precision of measurement and ease of adjustment were secured. It was in the United States that the best work in this field began to be carried on. The lenses of Alvan Clark gained an international reputation. An objective 30 inches in diameter was made by him for the Russian Observatory at Pulkova soon after a 26-inch telescope had been completed for the U. S. Naval Observatory in Washington. These were succeeded by the 36-inch instrument of the Lick Observatory and the 40-inch telescope of the Yerkes Observatory, with both of which results in proportion to their increased size have been obtained.

The seventeenth century really marks the beginning of instrumental work and accurate measurements in astronomy. The vernier, which made it possible to subdivide linear and circular scales with accuracy, made its appearance in 1631. In 1640 the optical axis or line of direction of the telescope was practically defined, and the micrometer was invented by William Gascoigne (1612-1644), which was the forerunner of the filar micrometer, so essential to modern astronomy, where an image at the focus of a telescope can be measured.

The micrometer is indeed an important adjunct to the telescope, for, unless angular distances can be measured, the mere bringing nearer of the celestial bodies would have but a limited amount of usefulness. In the micrometer of William Gascoigne two pointers carried by a single screw were placed at the focus of a telescope. When these pointers were parallel they pointed to zero; but, by revolving the screw, they could be separated and the number of revolutions or parts of a revolution could be read from a divided head. Consequently all that it was necessary to know was the distance between two successive threads of the screw in order to obtain an exact value for any distance which the pointers might separate. Now if it were desirable to determine the angular distance between two stars, each pointer was set on a star and the distance between them was thus gradually measured, so that by simple mathematics the corresponding angular distance could be computed.

Micrometers soon became an important part of exact observation with a telescope. Auzout and Picard made subsequent improvements, so that finally a micrometer resulted in which a spider filament was placed on a frame moved by a screw with graduated head, thus enabling increased precision of observation to be obtained. This is the fundamental device now used with various improvements and refinements. Roemer, who was the first to determine the velocity of light, improved the micrometer in 1672 by adding springs to take up the lost motion. He also constructed the first meridian telescope in 1689. By the middle of the seventeenth century the use of telescopic sights for determining the position of the stars had become established. The precision of the observations of that epoch may be estimated at 10 seconds of arc, which corresponds to the diameter of a lead pencil seen at a distance of about 550 feet.

Methods and instruments continued to improve. The observations of Lalande attained a limit of precision of one second of arc, corresponding to a pencil at 5,500 feet. At the beginning of the nineteenth century great improvements were made. In 1875 the limit of precision had been reduced to one-half a second, which removes the lead pencil to 11,000 feet or more than 2 miles.

For minute measurements one of the most useful devices has been the heliometer or divided object glass micrometer, the first really available type of which was constructed by Fraunhofer for the observatory at Königsberg. In this instrument an object glass or lens is used which is divided along its diameter. The two parts of the glass are mounted so that they can be moved laterally with respect to each other. Consequently each half supplies a distinct image of the same object, but separated by a strictly measurable amount. Thus, if a double star is under examination, the two half lenses into which the object glass is divided can be moved until the upper star in one image is brought into coincidence with the other star in the lower iamge, so that the distance apart becomes known by the amount of motion employed. By using screws with heads of considerable size to move the halves of the object glass, the heliometer can be read to the thousandth part of a revolution, and in the case of the Königsberg instrument such a division, equivalent to $\frac{1}{20}$ of a second of arc, could be measured with accuracy. This new instrument, which was not mounted until 1829, three years after the death of Fraunhofer, was at once employed by Bessel to solve the problem of star distances. His measurement of the parallax of the star known as "61 Cygni," corresponding with a distance about 600 times that of the Earth from the Sun, not only was considered ascertained beyond question, but is spoken of by Miss Clerke as "memorable as the first published instance of the fathom line so industriously thrown into celestial space having really and indubitably touched bottom." In 1874 the heliometer was applied to the observation of the transit of Venus, and again in 1877, when Mars came into opposition with the Sun, Sir David Gill, using the heliometer, made a valuable determination of the solar parallax, obtaining a value of 8.78 seconds, corresponding with a distance of 93,080,000 miles. By this time the heliometer had become an accepted method for improving astronomers' knowledge of the Sun's distance. A number of heliometers were employed in coöperation at different points of the Earth's surface, the work of Professor Elkins at Yale in connection with Sir David Gill at the Cape of Good Hope Observatory being especially notable.

Another modern development of telescopic astronomy has been the direct measurement of the magnitude and brightness of a star, thus superseding to a great degree the judgment of the eye upon which the older astronomers had depended from the days of Hipparchus. The photometers (light measurers) used with telescopes for this purpose consist either of those designed to cut down the amount of light furnished by a measurable amount and thus cause the star to disappear, or those in which conditions are so arranged that the light of the star appears just equal to some standard light. Under the first head is the so-called "cat's eye," in which a wedge of dark neutral tinted glass is placed close to the eye, either at the eyeball of the eyepiece or at the principal focus where the micrometer wires are usually placed. As the wedge is introduced until the star just disappears the graduation is read, which graduation can be reduced to a scale of magnitudes. In the other class of photometers the light of the star is compared with an arbitrary artificial star formed by light from an oil lamp shining through a small aperture.

To Huygens is due the application in 1655 of the pendulum to the practical measurement of time, thus giving us a clock so regulated that it was possible to make accurate time observations. The invention of the pendulum clock, patented in 1657, therefore marks a distinct epoch in astronomy.

The most usual and most useful form of mounting for a telescope is the equatorial, the principal axis of which is inclined at an angle equal to the latitude of the observatory and is directed toward the North Pole in the Northern Hemisphere and toward the South Pole in the Southern Hemisphere. The axis of the instrument is thus parallel to the Earth's axis of rotation and is therefore called the polar axis. It carries a graduated circle which is parallel to the celestial equator, known as the hour circle, from which circle may be read the hour angle of the body upon which the telescope happens to be pointed. The polar axis also carries the bearings of the declination axis, which is perpendicular to the polar axis and carries the telescope itself and the declination circle. When the equatorial telescope is directed toward a star or a planet it is necessary only to use clock-work machinery to cause the polar axis of the instrument to turn with a uniform motion in order to follow any star or planet which otherwise would soon be carried out of the field of view by the rotation of the Earth. The equatorial also enables the observer to look at once to a particular part of the heavens where a given body is expected to be at a given time.

The mounting for a modern equatorial telescope requires large and heavy moving parts. Where a solar or stellar image is desired it does not seem desirable to employ such a heavy mechanism. To Léon Foucault, about 1868, the idea occurred to construct a fixed telescope with a mirror, moving with one-half the angular velocity of the Sun, deflecting a beam in a fixed direction. Such an instrument was constructed and was employed with good results, altho its operation was marred by imperfections in its driving mechanism. However, the device did not attract much attention until plans were made in the eclipse expedition of 1890 for extensive photographs of the phenomenon. It was proposed to use the instrument in connection with a second mirror to produce an image which would not move. This device, now called a "cœlostat," was found admirable for eclipse photography. Experiments were made at the Yerkes Observatory to construct such an instrument for solar work. The work was subsequently transferred to the Carnegie Institution Observatory on Mount Wilson. An extension of the same principle may be found in the tower telescope of that institution, where a cœlostat is mounted on the top of a skeleton tower and a beam of light is reflected to a laboratory beneath. To-day the most modern and efficient reflecting telescope of large size is the 100-inch instrument designed for the same observatory by Prof. G. W. Ritchey.

At the present time both refracting and reflecting telescopes are in use and have been brought to a great degree of perfection. Just which is the better it would be hard to say. The old speculum metal reflector has been almost discarded and glass, coated with silver, has been substituted. The glass is much superior to the metal, as it can be figured more accurately, and if tarnished the silver can be removed without changing the figure of the mirror.

Again, much study has been given in France to a form of telescope known as the equatorial coudé, in which the optical axis of the telescope is parallel to the axis of the Earth and the light of the star is reflected into it by two mirrors. Such an instrument, constructed for the Paris Observatory, has been very convenient for the astronomer, who can sit in his chair and observe the stars as easily as he can use his microscope. But the loss of light and definition by the double reflection, as well as the deflection of the mirrors and the varying temperatures to which the different parts of the instrument are subjected, render this construction far from perfect.

CHAPTER III

THE EVOLUTION OF ASTRONOMICAL INSTRUMENTS AND METHODS—THE RISE OF ASTROPHYSICS—THE SPECTROSCOPE AND ITS MODIFICATIONS

To GAIN a knowledge of the composition and nature of the celestial bodies is the fundamental probl m of astronomy. Unable to bring a celestial body or a specimen from it, except in the rare case of a meteorite, to the chemical laboratory for study, the astronomer is dependent entirely on a study of the energy that it emits in the form of light and heat rays. Strange as it may seem, these rays furnish as true a record of their birth and life history as if a sample from the distant star had been tested in the assay furnace or with the reagents of the chemist. The simple instrument called a spectroscope gives an accurate and permanent record which affords complete data for the studies of the astronomer.

White light is composed of various forms of vibration which, taken by themselves, will supply light of various colors from red to violet. It was found by Sir Isaac Newton in passing a beam of white light through a prism that not all of the rays are bent equally toward and away from the perpendicular, but that the amount of bending depended upon their color, or as it is now termed, their wave length and position in the solar spectrum. Thus, when he permitted a beam of white light, emerging from a hole in a shutter, to fall upon a prism in a dark room, he found that there was produced after its emergence a brightly colored band with the red at one end, where the waves were refracted the least, and shading through yellow and green to violet, where the waves were bent or refracted most. Consequently, if there were a source of light capable of furnishing one color and that only, it would be obvious, wherever that color appears in the bright band produced by the prism, that it radiated from a particular source.

Before 1753 a young Scotchman, Thomas Melvill, noticed that when various compounds of sodium were introduced into an alcohol flame and viewed through a prism

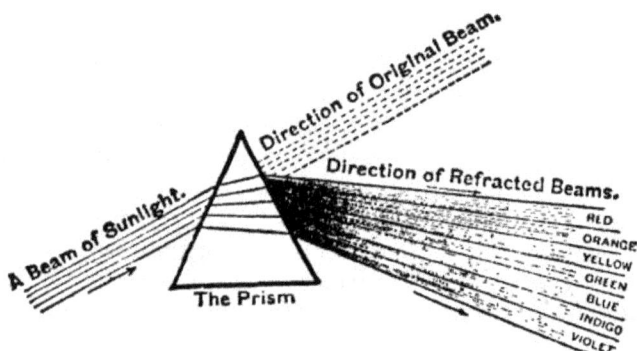

Fig. 3 —Dispersion of Light by the Prism.

there appeared a particular shade of yellow light, which was always bent or refracted to a fixed and definite degree. Others repeated these experiments, and finally Fraunhofer (1787-1826), a great optician of Munich, rediscovered this deep yellow ray and found its place in the spectrum. The same phenomenon was noticed by many other experimenters. Indeed, the omnipresence of the yellow light was often an embarrassment in spectral research. That this yellow line was due to sodium was pointed out by William Swan. Finally, it was noted that the distribution of sodium was so general and the prism test of its presence so delicate that its absolute exclusion was well nigh impossible.

Before Fraunhofer's experiment., the round hole in the
shutter of Newton had been supplanted by a slit or crevice
about one-twentieth of an inch wide, and the spectral band
thus formed from sunlight was not only continuous but
free from overlapping images, so that the colors were
shown in their purity, crossed by seven dark lines.   In
the course of his experiments Fraunhofer not only used
the slit, but added to it the telescope of the modern spectroscope.   He was surprised to find not merely seven but

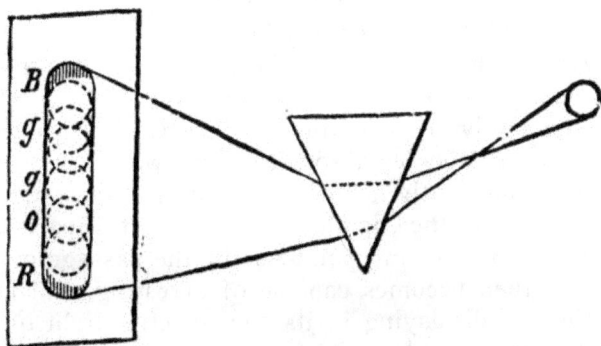

Fig. 4 —NEWTON'S DIAGRAM.
Showing spectral band from blue (B.) to red (R.) formed by a
prism from a beam of sunlight coming from the round hole
in a shutter of a darkened room.

thousands of dark transverse lines, many of which he
mapped, counted and designated by the letters of the alphabet.

Not only did he examine sunlight in this way, but also
the light of the Moon and planets, and found that stellar
spectra, too, were crossed by the same dark lines.   In the
case of certain stars there were even dark bands.   He
found that one or rather a pair of solar lines which he
had marked in his map with the letter "D" coincided exactly with the yellow beam which accompanied incandescent sodium vapor.   The coincidence was noted by Fraunhofer, but the explanation came in 1859 from the distin-

guished German physicist, Professor Gustav Robert Kirchoff (1824-1887). He it was who sent a beam of bright sunshine through sodium vapor and discovered that the "D" line of Fraunhofer, instead of being effaced by the flame, was strengthened. The same held true with iron. The inference was of course drawn that sodium and iron were constituents of the glowing atmosphere of the Sun and that light of the particular wave length in passing through such an atmosphere was absorbed.

This principle has been formulated by Miss Clerke as follows: "Substances of every kind are opaque to the precise rays which they emit at the same temperature—that is to say, they stop the kinds of light or heat which they are then actually in a condition to radiate. But it does not follow that cool bodies absorb the rays which they would give out if sufficiently heated. Hydrogen at ordinary temperature, for instance, is almost perfectly transparent, but if raised to the glowing point—as by the passage of electricity—it then becomes capable of arresting, and at the same time of displaying in its own spectra, light of four distinct colors." In these few words we have the essence of spectroscopic chemistry and astrophysics. Materials of the Earth when heated to incandescence give a bright line spectrum characteristic of the individual element, but the same materials in the Sun show a spectrum marked by dark lines.

While spectrum analysis was applied to chemistry and terrestrial materials by Bunsen, Kirchoff worked industriously and made a map of the solar spectrum some eight feet in length, in which the various lines of the elementary bodies were represented. The spectroscope, as constructed by Kirchoff, consisted of a slit placed at the principal focus of the convex lens to make the rays parallel for their passage through the prism. In order to secure greater dispersion, several prisms were added and the emerging beam was passed through the telescope to form an image of the spectrum. The new instrument at once presented an enormous number of lines for study, not only of the Sun, but of various other celestial bodies It was soon applied to the observing end of an astronomical telescope, so that the celestial image was formed directly at the slit of the spectroscope.

By increasing the number of prisms the dispersion of the spectroscope can be increased, and a longer spectral band produced, in which otherwise closely adjacent lines are increasingly separated. But in passing through a number of prisms there is considerable loss of light by

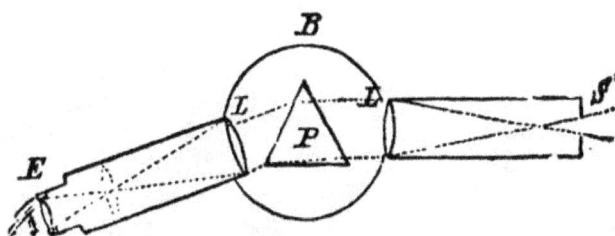

Fig. 5 —S., slit of collimating telescope directed toward source of light, lens of collimating telescope to render parallel rays from S.; P., prism ; B., base on which prism stands ; L., lens of observing telescope ; E., eyepiece of observing telescope.

reflection and absorption, so that a limit is soon set to the number of prisms employed. Another form of spectroscope, which makes use of a grating or a number of fine lines ruled very closely together on a transparent or a reflecting surface, has been found to possess greater dissolving power without any accompanying loss of light. In fact, the resolving power of a perfect grating depends simply upon the total number of lines it contains, so that the light efficiency per unit area may be as great for a large grating as for a small one.

The principle of the grating depends upon the interference of the various minute light waves caused by a series of lines, amounting to from ten to twenty thousand to the inch, ruled on a transparent or a reflecting surface. The mathematical discussion of the formation of the spectra by the interference of the light waves in passing through or being reflected from such a grating can hardly find a place here. The result, however, is essentially the same as in the case of the prism. As soon as the dispersion of light was obtained by this means it was found that it could be studied quantitatively, and that the grating could be used for

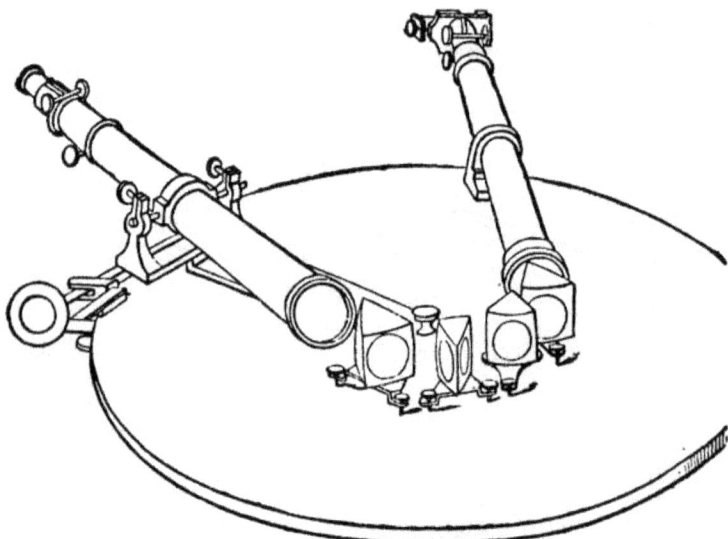

Fig. 6 —KIRCHOFF'S SPECTROSCOPE.

astronomical measurements with as great facility as the prism spectroscope.

Lewis M. Rutherfurd of New York was able to make excellent gratings about 1864, but it remained for Professor Henry A. Rowland (1848-1901) at the Johns Hopkins University to construct a dividing engine with a screw, practically free from error, which would move a small plate of polished speculum metal by regular intervals of $1/_{15000}$ of an inch under a diamond point which traced sharp and regular lines. This machine not only was remarkably sensitive in its action, but automatically compensated for any minute irregularities in the screw. It was made to work at a constant temperature. It automatically proceeded with its ruling night and day until a grating of the desired length was completed. Professor Rowland for the first time ruled gratings on concave surfaces and used them in place of the prism of the ordinary spectroscope. The spectra obtained with these diffraction gratings in conjunction with special lenses were many feet in length and could be photographed in sections on photographic plates, each of about 20 inches in length.

The grating spectroscope has been modified by Professor A. A. Michelson of the University of Chicago, who has devised a new form of grating in which a series of glass plates precisely equal in thickness are placed one on another like a flight of steps. A parallel beam of light when transmitted through them is resolved into spectra of a very high order, exceeding even those of Rowland's largest gratings, so that compound lines in the spectrum can be studied with facility.

The application of the spectroscope has made of astronomy an experimental science, with methods and instruments for research and future progress fully as promising as may be found in any of the physical or natural sciences. The spectroscope has not only amplified astronomy, but it has developed the new science of astrophysics, in which astronomy is combined with physics. New methods and instruments for research already have brought to light striking discoveries which have compelled the modification of older astronomical and cosmical theories.

In connection with the spectroscope it is possible to measure the temperature of the radiations sent out from the Sun and the stars with a high degree of accuracy by means of the bolometer, a sensitive thermometer, invented by Professor S. P. Langley. It consists of two very fine threads of platinum wire about $1/2500$ of an inch in thickness, mounted side by side within a constant temperature chamber. On one of these wires the radiation is permitted to fall, while the other is carefully shielded. Any change in the temperature of the wire on which the light or heat waves fall produces a difference in its electrical resistance that can be measured with a high degree of precision, so that a difference of less than one-millionth of a degree in the temperature can be clearly indicated.

The spectrum formed by the spectroscope is caused to move slowly across the exposed platinum wire of the bolometer and a galvanometer in the circuit reflects from its mirror a spot of light upon a photographic plate, so that the deflections of the magnetic needle are photographed and registered, thus indicating the intensity and energy at the different parts of the spectrum. This instrument was first used by Langley to determine the amount of heat received from the Sun on the top of Mount Whitney in 1881, and since that time it has been employed by him and his successors at the Astrophysical Laboratory at Washington and also at Mount Wilson. The problem that the bolometer seems capable of solving is to determine the atmospheric absorption of light and heat in the passage of the Sun's rays to the Earth. It has also been used to measure the heat spectrum of the Moon and some of the brighter stars; in the case of the former showing that the Moon is very cold, as there is a considerable quantity of heat radiated having a wave length greater than that of the heat radiated from a block of ice.

After the fundamental work of Kirchoff in identifying the spectral lines of the Sun and the stars with various terrestrial materials it was but natural that the composition of stars as shown in their spectra should be thoroly attacked by astronomers. Among the first of these was Sir William Huggins, who devoted the greater part of a useful scientific life to research of the heavenly bodies, especially as revealed by the spectroscope.

In 1862 Huggins, Secchi and Lewis M. Rutherfurd began their researches in stellar spectra that enabled them to classify and compare the spectral bands furnished by the different stars. It was the spectroscope in the hands of Sir William Huggins that made possible the solution of the riddle of the nebulæ, the nature of which for long years had been a vital point of discussion among astronomers. On August 29, 1864, directing his spectroscope to the planetary nebulæ in Draco, Huggins saw, instead of the bright band he anticipated, a single line which subsequently was resolved into three lines. Thus he proved that the nebula was not an aggregation of stars or incandescent solid materials, which would have afforded a continuous spectrum crossed by dark bands and a luminous gas.

The effect of the spectroscope on astronomical research is thus summarized by Professor Hale: "In astronomy the introduction of physical methods has revolutionized the observatory, transforming it from a simple observing station into a laboratory. The interest of the student of astrophysics is no longer confined simply to celestial phenomena. For astrophysics has become, in its most modern aspect, almost an experimental science, in which some of the fundamental problems of physics and chemistry may find their solution. The stars may be regarded as enormous crucibles, in some of which terrestrial elements are subjected to temperatures and pressures far transcending those obtainable by artificial means. In the Sun, which appears to us not merely as a point of light like the stars, but as a vast globe whose every detail can be studied in its relationship to the general problem of the solar constitution, the immense scale of the phenomena always open to observation, the rapidity of the changes and the enormous masses of material involved provide the means for researches which could never be undertaken in terrestrial laboratories. Hence it is that astrophysics may equally well be regarded as a branch of physics or as a branch of astronomy."

The great advantage of the spectroscope over the eye or the direct image from the photographic plate is its ability to analyze the action of light. While the intensity of light suffers in its journey through space, yet the nature or character of the light undergoes practically no change, so that the light from a distant star, separated from the Earth by an interval that seems to us almost infinite, can be received in our spectroscope and be resolved into a spectral band with difficulty but slightly greater than that which would be found in the employment of a luminous source of light at the opposite end of the laboratory table.

The spectroscope, therefore, can be used in astronomy to determine the composition of a distant body according to the principles of spectrum analysis. But this is not all. It also enables the determination whether the light from a luminous body in the heavens is approaching or receding, and whether the light emitted from such a body is the same to-day as it was yesterday or a half century ago, and whether it comes from one or more bodies which the eye and perhaps the telescope cannot separate, but which are distinctly separate. Hence the astronomer is only too glad to remove the eyepiece from his telescope and put in its place some spectroscopic device which will analyze the light into separate colors and give him much valuable information as to the constitution and motions of even the most distant star.

The value of the spectroscope is greatly increased by the application of photography. The general nature of a spectroscopic investigation can best be indicated by abstracting from Professor Hale his description of solar spectrum analysis: "Sunlight must be reflected from a mirror to a heliostat (driven by clockwork, to maintain the beam in a fixed direction) to the slit. Between the slit and the heliostat a lens is introduced, for the purpose of forming an image of the Sun upon the slit. When the illumination is secured in this way, the whole grating is filled with light from the diverging rays. The grating then produces an image of the solar spectrum upon the photographic plate, where it may be recorded by giving a suitable exposure.

"To facilitate an accurate comparison, the solar spectrum is photographed side by side on the same plate with the spectrum of the substance whose presence in the Sun is to be determined. In order to accomplish this, one-half of the slit is covered, and the sunlight is admitted through the other half. Thus the solar spectrum is photographed on one side of the plate. After this exposure is completed, the sunlight is shut off, and the screen in front of the slit moved so as to cover the half previously open and to uncover the other half. The image of the Sun on the slit of the spectroscope is then replaced by an image of an electric arc light, burning between two poles of iron. The spectrum of the iron vapor is thus produced on the plate, and a strip of this spectrum is photographed beside the strip of solar spectrum.

"The bright lines of iron are represented in the solar spectrum by corresponding dark lines which accurately image them in position. In Rowland's work on the solar spectra thousands of lines were found to correspond with the iron lines given by the electric arc.

"The same process can be employed to determine the presence of other substances in the Sun. In the case of metals, the electric discharge may be caused to pass between two metallic rods, or fragments of the metal may be placed in a hole drilled in one of the carbons of an ordinary electric arc-lamp. In the latter case the spectrum of carbon, and also of impurities which the carbon poles always contain, will appear on the plate with the spectrum of the metal in question. But these extra lines may always be identified, and usually give no trouble. The identification of the solar lines, however, is not always so simple as in the case of iron. Many substances are doubtfully represented in the Sun by only a small number of lines, and it is sometimes very difficult to decide whether a few apparent coincidences are sufficient to warrant one in drawing definite conclusions. The matter is usually determined by ascertaining whether certain well-known groups of lines, which for various reasons are considered to be especially characteristic of an element, are actually represented. If these groups are absent, an apparent coincidence with certain less characteristic lines belonging to the same element should be regarded with suspicion. In the case of gases, the comparison is effected by the aid of vacuum tubes, in which the gas, usually at low pressure, is illuminated by an electric discharge. Thus the lines given by a hydrogen tube in the laboratory have been shown to coincide in position with lines ascribed to hydrogen in the Sun.

"After many years of study of the solar spectrum by these methods Rowland reached the conclusion that the chemical composition of the Sun closely resembles that of the Earth. Certain elements, such as gold and radium, iodine, sulphur and phosphorus, chlorine and nitrogen, have not been detected in the Sun. But this does not prove that they are certainly absent, as their level in the solar atmosphere, or the low degree of their absorptive effects, might prevent them from being represented. On the other hand, various substances not yet found on the Earth are shown by many unidentified lines of the solar spectrum to be present in the Sun. Some if not all of these will probably be discovered by chemists, just as helium was found by Ramsay in clevite. Rowland remarked that if the Earth were heated to a sufficiently high temperature it would give a spectrum closely resembling that of the Sun."

CHAPTER IV

AFTER the spectroscope, photography has been the most useful tool of the astronomer, and to its aid must be credited some of the most important work of the latter half of the nineteenth century. For the development of celestial photography as outlined in the present chapter an interesting paper by Professor E. E. Barnard supplies in large part the material. According to Professor Barnard, the application of photography to astronomy may be said to date from the very first announcement of Daguerre's wonderful discovery of the production of a permanent image by the effect of light upon silver salts. "The celebrated French astronomer, Arago, quickly foresaw its great possibilities, especially in the faithful delineation of the surface features of the Sun and the Moon, for these two objects at least were bright enough to register themselves with the sluggish materials then in use." It was of course obvious to astronomers and physicists that the formation of an image on a sensitized plate was in no way different from that produced at its focus by the telescope lens and that the image of a celestial body could be produced as well as any other.

It was from America that the first practical work came, and "within less than one year from the announcement of Daguerre's discovery, in March, 1840, Dr. John W. Draper of New York city succeeded in getting pictures of the

Moon which, though not very good, foreshadowed the possibilities of lunar photography. Five years later, at the Harvard College Observatory, Bond, with the aid of Messrs. Whipple and Black, of Boston, succeeded in obtaining still better pictures of the Moon with the 15-inch refractor. These pictures on daguerreotype plates aroused great interest, especially in England. However, the difficulties encountered led to failures generally, except in the case of De la Rue, Dancer and one or two others. In 1858 De la Rue, using a 13-inch metal speculum reflecting telescope, without clockwork, and guiding it by following a lunar crater seen through a plate, made the most important of the early efforts at lunar photography." His photographs were the best until those made in America, in 1860, by Dr. Henry Draper, son of the illustrious John W. Draper. He secured excellent photographs of the Moon, superior to any previously made, and capable of considerable enlargement. These pictures were the best taken until Lewis M. Rutherfurd began his remarkable work about 1865. His admirable photographs of the Moon were made with a refractor of 11-inch aperture, which, constructed under his immediate supervision, was the first telescope corrected especially for the photographic rays.

"The completion of the Lick Observatory in 1888 marked another decided advance in astronomical photography, especially of the Moon. The great focal length of this magnificent instrument gave an unenlarged image of the Moon about six inches in diameter, which in itself was a great advantage." Good results were also secured with the Yerkes refractor.

Admirable lunar photographs have been made by MM. Loewy and Puiseux, with the equatorial coudé, at Paris, and have shown the usefulness of this singular instrument for such work.

"The first picture of the Sun seems to have been made on a daguerreotype plate by Fizeau and Foucault in 1845," says Professor Barnard. "During the total eclipse of

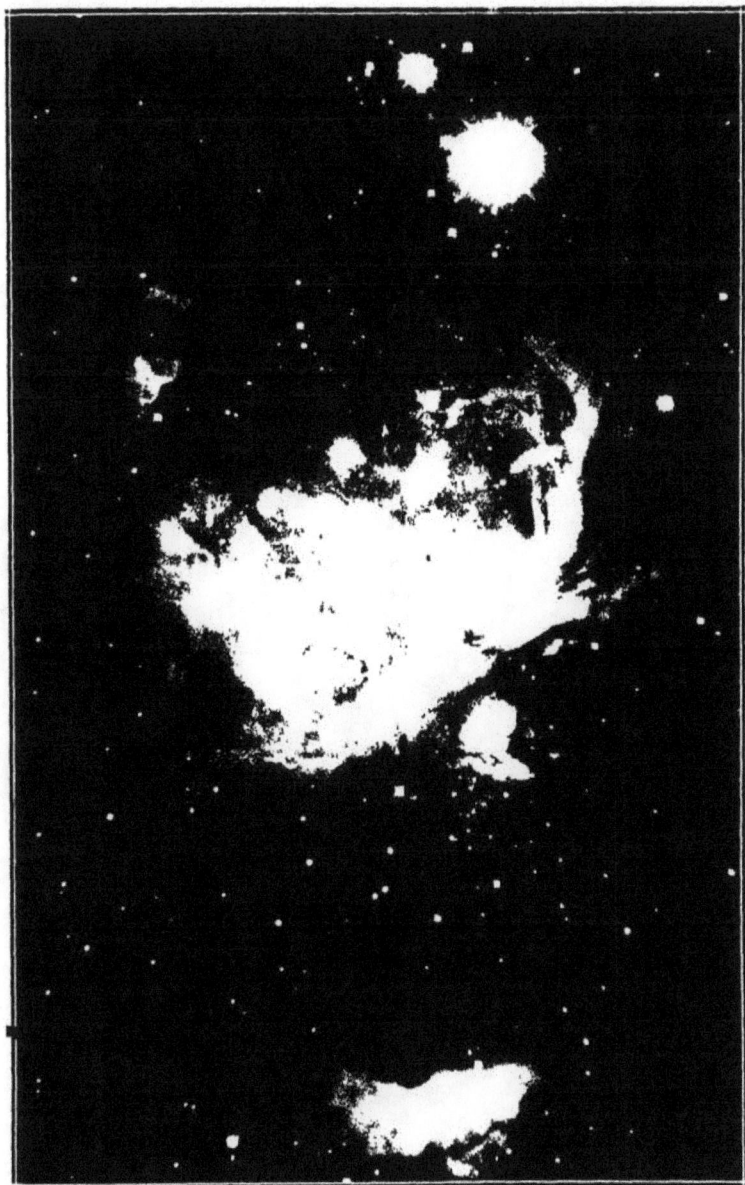

THE GREAT NEBULA IN ORION.

the Sun on July 28, 1851, a deguerreotype was secured
with the Königsberg heliometer (2.4 inches in diameter
and 2 feet focus) by Dr. Busch, which appears to have
been the first photographic representation of the corona.
It showed considerable detail quite close to the Moon."

But in the early eclipses photographic work seems to
have been devoted mainly to representations of the solar prominences, which at that time were as rarely seen as
the corona itself. "During the eclipse of 1869, however,
Professor Himes secured a photograph which showed the
brighter structure of the corona. Similar pictures were
also obtained during the same eclipse by Mr. Whipple, of
Boston. The corona was also slightly shown on pictures
made as early as 1860 by M. Serrat. None of them, however, showed more than slight traces of the corona, extending only for a few minutes of arc from the Moon's
limb. Nearly all the pictures seem to have been taken
with an enlarging lens, which was doubtless used to get
the prominences on a larger scale."

"The first really successful photographs of the corona
were obtained at the eclipse of December 22, 1870, when it
was shown on the plate to a distance of about half a degree from the Moon's limb. This picture, made by Mr.
Brothers, at Syracuse, Sicily, showed a considerable
amount of rich detail in the coronal structure; the same
can also be said of the photographs of this eclipse taken by
Colonel Tennant and Lord Lindsay's party. These seem
to have been the first pictures that really showed the great
value of photography for coronal delineation. The eclipse
of 1871 was still more successfully photographed, and an
excellent representation of the corona, full of beautiful
•detail, was secured."

"In 1878 extensive preparations were made to observe
the eclipse of July 29 of that year. Photography was to play
an important part, though astronomers did not rely very
strongly upon it; for it appears that all were prepared to
make the customary drawings of the corona. Unfortu-

nately each person faithfully carried out that purpose. A most suggestive illustration of the uncertainty of such work is found in the large collection of drawings published in a volume issued by the United States Government relating to the eclipse of 1878. An examination of these forty or fifty pictures shows that scarcely any of them would be supposed to represent the same object, and none of them at all closely resembled the photographs. The method of free-hand drawing of the corona made under the attending conditions of a total eclipse received its death blow at that time, for it showed the utter inability of the average astronomer to sketch or draw what he really saw under such circumstances."

In the eclipses of 1882, 1886 and 1889 photography played a part of increasing importance in the observations. In the latter year there were a large number of amateur photographers who took advantage of the eclipse to make many photographs, which, in a number of cases, were taken in a systematic and scientific manner. At the Lick Observatory a beginning was made in eclipse photography with an extemporized apparatus and successful exposures were made. During the eclipse of 1896 important work was done in photographing the flash spectrum or the momentary reversal of the Fraunhofer lines which occurs when the edge of the Sun disappears behind the Moon or reappears from it and for an instant exposes the reversing layer, which was first seen by Professor Young at the eclipse of 1870. This photograph was made by a young Englishman, William Shackleton, who, on exposing a plate at the critical instant of the reversal of the lines, caught for the first time the fugitive bright lines which are visible for only about a minute. This gave a permanent visible record of the phenomenon which removes it from the class of hasty visual observations, whose results depend upon the memory of the observer.

The photographing of such a minute point of light as a star is quite different from that of a luminous or brilliant body like the Sun or Moon. Yet it was early essayed, and from the first photograph of a star by Bond in 1850 to the present time stellar photography has gradually risen to a prominence as remarkable as it is important. Indeed, it is now quite indispensable. The principal reason for the real increase of importance in this work, however, was the succesful introduction of the very rapid dry plate. The wet or collodion process, which astronomers soon pushed to its limits, was poorly adapted to the photography of the stars, and of no use whatever for comets and nebulæ. "Notwithstanding the inherent difficulties of the wet plate, the photographs of the star clusters, etc., of the southern skies, obtained under the direction of Gould with an 11-inch photographic refractor by the wet process, were of the highest value and showed upon measurement a striking agreement in accuracy with visual work. The same can be said of Rutherfurd's photographs of the Pleiades, Praesepe, etc., which were made prior to Dr. Gould's, and which were the first photographs of this kind."

"As early as 1857 Bond had shown, by measurement of a series of photographs of the double star Mizar, that the highest confidence could be placed in measures of star plates. This was subsequently fully verified by Gill, Elkin and others. As regards absolute accuracy Dr. Elkin showed in 1889 that measures of a photograph of the Pleiades taken by Mr. Burnham, with the great telescope at Mount Hamilton, had equal value with the heliometer measures of the same stars."

By 1881 or 1882, however, dry or gelatine emulsion plates were beginning to be used with every promise of their ultimate value, as was shown by the photographs of the comet of 1881, which were made by Draper and Janssen. These were the first photographs ever made of a comet. Efforts had been made to secure pictures of Donati's comet in 1858, but without success.

It was quite obvious that as soon as satisfactory photographs of the stars were secured some earnest effort would be made to make use of them in a quantitative and systematic way. Previously, for the production of star maps and catalogues, elaborate series of observations were made at the various observatories and the positions of the stars computed and incorporated in large volumes. At the Royal Observatory at the Cape of Good Hope Sir David Gill, in 1882, after making some pictures with a large camera of the comet of that year, found that not only did the plate show the stars visible to the naked eye, but a number as small as the ninth or tenth magnitude. Accordingly it occurred to him that such photographs furnished a novel and excellent method of cataloguing the stars and mapping the heavens, as it was necessary only to measure on the glass negatives the positions of the various stars and refer them to certain well-known points of reference. From 1887 to 1891 the entire southern heavens from 18° south declination to the celestial pole were duly photographed. The half million stars found on the negatives were then measured and the magnitude of each determined by Professor J. C. Kapteyn at the University of Groningen, Holland. Thus in 1899 was finished the Cape photographic "Durchmusterung," which is published in three quarto volumes and contains the magnitude and approximate position of every star photographed, the magnitude of the stars on each plate being reduced to a visual scale.

At the time when Sir David Gill began his photographic work, Dr. Barnard states, "the Henry brothers of Paris were making a chart of the stars along the ecliptic in their search for planetoids. They had at this time reached the region of the Milky Way, and the marvelous wealth of stars they encountered on entering the boundaries of that vast zone completely discouraged them from carrying their charts through the rich region traversed by the ecliptic. While hesitating as to the advisability of continuing their work, the photographs of the great comet came to their notice. They were struck with the great number of stars shown on these pictures together with the image of the comet. The idea at once occurred to them that they could use this wonderful process to make their charts. They began at once the construction, with their own hands, of a suitable photographic telescope of 13½ inches diameter for the photography of the stars. This instrument produced exquisite star pictures, which were marvels of definition, as well as photographs of the nebulæ, of Saturn and Jupiter, the Moon, etc."

It was the success of the Henry brothers' work that led to the International Astro-Photographic Congress, which met at Paris in 1886. This Congress undertook the organization of an International Commission engaged in the preparation of a photographic chart and catalogue of the heavens, and the work since that time has been actively in progress. Uniform instruments of the same aperture and focal length are used at the eighteen observatories participating in this work and two sets of plates are being made, one to include all the stars that are capable of being photographed and the other one those of the eleventh magnitude. With this photographic map astronomers anywhere can compile their own catalogues, and portions of such catalogues by various national observatories have already been issued. The method of preparing the chart consists in photographing the whole sky upon glass plates about 8 inches square. Each observatory has had assigned to it definitely its part of the sky, and about 11,000 plates of the size specified will be required to complete the task. Each plate of course carries one or more well determined catalogue stars, whose position is known with accuracy, so that from such points of reference it is possible to determine exactly the position of any other star on the plate.

"The photographic plate not only did away with the necessity of making the star charts by eye and hand, so essential to facilitate the discovery of planetoids, but it also did away with the necessity of the charts themselves for that purpose. The little planet, which is moving among the stars, now registers its own discovery by leaving a short trail—its path during the exposure—on the photographic plate. The first of these photographic discoveries of planetoids was made by Dr. Max Wolf in 1892, and his observatory at Heidelberg subsequently became a headquarters for discoveries of this kind. Planetoids are now found wholesale in this manner by photography."

In the early days of photography nebulæ were considered the most unpromising subject for the photographic plate to deal with. Most of these objects appeared so faint that but little encouragement in that direction was offered the celestial photographer.

"One of the brightest and most promising of nebulæ is that in the sword of Orion, and this was naturally one of the first of these objects to receive photographic attention. In September, 1880, Dr. Henry Draper began photographing nebulæ with this object, and succeeded, with 51 minutes exposure, in getting a good picture of the brighter portions on dry plates. This was the first nebular photograph. It was followed by other photographs, one of which showed stars down to the 14.7 magnitude which were visually beyond the reach of the same telescope. These pictures marked a new era in the study of nebulæ. When the results were communicated to the French Academy by Dr. Draper, Janssen took up the work with a reflecting telescope having a silver-on-glass mirror of very short focus, constructed in 1870 for the total solar eclipse of 1871. With this Janssen found it easy to photograph the brightest parts of a nebula with comparatively short exposures. Unfortunately for science, the death of Dr. Draper, in 1882, put a stop in America to the work he had inaugurated, but it was at once taken up in England by Common, who, with a 3-foot reflector, attained rapid and immediate success. His photographs of the great nebula of Orion are still classic. They were a great advance over the work of Draper, for the reflector was not only a larger telescope, but was also better adapted for photographic purposes, and especially for photographing nebulæ. In fact, as we shall see in a later chapter on nebulæ, much of the progress in their study has been due to photography.'

The photography of nebulæ was carried on with remarkable success at Lick Observatory during the incumbency of Professor James E. Keeler as Director. Using the Crossley reflecting telescope, presented to the Observatory by Dr. Common, he made a photographic study of nebulæ, and reached the conclusion that there are at least 120,000 of the spiral type within the range of this instrument. Professor Perrine, who succeeded to this work on the death of Professor Keeler, believes that half a million is nearer the figure, and that with more sensitive photographic plates and longer exposures the number of spirals would exceed a million.

Not only stellar motion, but stellar distances, can be measured by photography. Professor Pritchard, at Oxford, has used the sensitive plate to sound the celestial depths. His first experiments were undertaken with the star 61 Cygni, and by measuring 200 negatives which had been made in 1886 he derived for that star a parallax of 0.438", which was in satisfactory agreement with Ball's value of 0.468". This work was subjected to detailed scrutiny, and the Astronomer Royal was convinced that it was more accurate than that of Bessel's results, obtained with the heliometer. This was the beginning of the method of measuring a parallax from photographic plates. Professor Kapteyn showed in 1889 that from such plates, exposed at desired intervals, parallaxes could be derived wholesale. He applied his system in 1900 to a group of 248 stars with encouraging success. In fact, it was suggested that a photographic parallax "Durchmusterung" should be undertaken after the completion of the astrographical chart of the heavens.

When used in connection with the spectroscope the photographic plate has a field singularly suited to display its possibilities. Here it deals not alone with what can be seen, but it enters into regions where the eye takes no cognizance of things. For tho it is partly blind to the light which affects the eye, it can readily penetrate the regions where man, in turn, is blind. By special treatment of the plate photography registers those rays invisible to the eye and permits their accurate measurement.

The spectograph, or combination of photographic apparatus with spectroscope, must be so arranged as to show with distinctness the greatest number of lines, the individual lines being separated; consequently there are various types of spectrograph, depending upon the purpose for which they are to be employed.

One of the combinations of the spectroscope with the photographic apparatus is found in Professor Hale's spectroheliograph, which consists of a spectroscope across whose slit the solar image moves at a uniform speed. Instead of the eyepiece there is a second slit which permits light from only a single line to pass and fall on the moving photographic plate, so that an image of the Sun, or a sun spot in light of a single wave length, can be made to fall upon the plate and thus be recorded.

The general effect of photography in astronomy may be summarized in the brief statement that it has removed the astronomer from the eyepiece of the telescope and has substituted the more sensitive photographic plate with its permanent record. "Hence it is that the present-day student of astrophysics does not correspond with the traditional idea of the astronomer," says Professor Hale. "His work at the telescope is largely confined to such tasks as keeping a star at the precise intersection of two cross-hairs, or on the narrow slit of a spectrograph, in order that stars and nebulæ, or their spectra, may be sharply recorded upon the photographic plate. His most interesting work is done, and most of his discoveries are made, when the plates have been developed and are subjected to long study under the microscope."

CHAPTER V

When the astronomers of old tried to account for the apparent motions of the heavenly bodies by complete systems of epicycles, they must surely have asked themselves, Why do the planets move so regularly? What makes them move thus? If they did, they troubled themselves but little to answer the inquiries in anything but a perfunctory way. For the most part they were content to regard the stars as the playthings of divinity, and the cause of their motions, therefore, as a mystery forever veiled to human eyes. Still one astronomer, Anaxagoras, did have some idea of a force which holds the planets in their orbits and which might be the same as that which operates upon substances at the surface of the Earth.

After his day (499 [?]-427 [?] B.C.) the idea seems not to have been expressed by any one until the awakening of science in the seventeenth century. Then Kepler darkly hinted at some attractive force, because his discovery of the mathematical curve described by the planets seemed to demand the existence of some constantly exerted controlling force and also because he had read Gilbert's 'De Magnete,' in which he was made acquainted with the phenomena of electrical attraction.

Such a force as he had in mind would act to maintain the motion of the planets and to drive them along in their orbits. But this was hardly the solution of the problem, since as Galileo found, the motion of a body of itself must

continue indefinitely, unless there is some cause at work
to alter or stop it. This formed the first and most important of the laws of motion which, if not independently discovered by Newton, were subsequently to be stated by him
with greater force and conciseness. The laws were of
primary importance, because they afforded a new and
correct way of considering not only the underlying reasons
for the motions of the planets but of all mechanical problems involving matter and motion.

Aside from the three great laws of planetary motion
established by Kepler as the result of many observations,
the most important lesson taught by him, and one that was
readily learned by Newton, was that the motions of the
planets were not to be attributed to the influence of mere
geometrical points, such as the centers of the old epicycles,
but to the actual presence of other bodies. Kepler suggested, in particular, that the planets might be considered
as connected with the Sun and therefore as sharing to
some extent the Sun's motion of revolution. From the
Sun emanated that special kind of influence which he assumed. Yet, while Kepler considered the Sun as the
source of this hypothetical force, he believed in a more
general gravity or attraction between bodies.

He was unfortunate enough, however, to conceive of it
diminishing simply in proportion to the distance between
the two bodies, a mathematical impossibility, as was demonstrated by Newton. This is the more surprising as he
had demonstrated that the intensity of light was recipro-
cally proportional to the surface over which it was spread
and that it varied inversely as the square of the distance
from the luminous body. It was also unfortunate that,
while Kepler's ideas of the nature of gravity were sound
and accurate in many respects, they bore no particular logi-
cal connection either one with another or with his theory of
planetary motion. They are, however, worthy of comment
as indicating the situation before Newton took these and
other speculative ideas and the three isolated laws of

planetary motion and bound them together into one beautiful doctrine which must underlie all astronomical science.

Kepler in his work, 'Commentaries on the Motions of Mars,' definitely states that gravity is a corporal affection, reciprocal between two bodies of the same kind, which tends, like the action of a magnet, to bring them together. When the Earth attracts a stone, the stone at the same time attracts the Earth, but by a force feebler in proportion as it contains a smaller quantity of matter. He then proceeds to state that if the Moon and the Earth were not retained in their respective orbits by an animal or other equipollent force, the Earth would mount toward the Moon one-fifty-fourth part of the interval which separates the two and the Moon would descend the fifty-three remaining parts, supposing it to have the same density. This idea of gravity, according to Kepler, was indeed general and served to explain the cause of the tides, as is clearly indicated in the following passage:

"If the Earth ceased to attract its waters, the whole sea would mount up and unite itself with the Moon. The sphere of the attracting force of the Moon extends even to the Earth and draws the waters toward the torrid zone, so that they rise to the point which is the Moon in the zenith."

After Kepler had promulgated his famous laws of planetary motion many minds independently conceived a force to account for the remarkable uniformity of that motion. Thus the idea occurred to Robert Hooke, to Christopher Wren and perhaps to Edmund Halley, who was Newton's most intimate friend and who probably did more than any other man of his time to popularize the idea of universal gravitation. It remained for the towering genius of Sir Isaac Newton (1643-1727) to formulate into a mathematical law of gravitation the effect of that universal force with which every schoolboy is now acquainted. The honor of having anticipated Newton was claimed by Hooke, and the two entered into an acrimonious controversy. Hooke never brought forward convincing proof of his claims. So far as Newton is concerned, the great merit of his work lay not so much in conceiving the law of gravitation as in his brilliant demonstration of its truth.

Starting with Kepler's laws of planetary motion, he showed not only that they were true, which was hardly a task of merit after Kepler had considered the observations of Tycho Brahe and all other astronomers whose recorded observations would throw any light on the subject, but why these laws were true, and why no other laws could have accounted for the conditions actually observed in the motion of the planets. And, furthermore, underlying these famous planetary laws he discovered must be the attraction of gravitation. By a mathematical analysis unrivaled in the history of astronomy he proved his theorem completely. Not only did he suggest, as did Kepler, that the power of attraction resided in the Sun, but he proved mathematically that as a necessary consequence of that attraction every planet must revolve in an elliptical orbit around the Sun, having the Sun as one focus; that the radius of the planet's orbit must sweep over equal areas in equal times and that in comparing the movements of two planets it is necessary that the squares of the periodic times be proportional to the cubes of the mean distances. These facts were discovered by Kepler; they were explained by Newton, with the aid of the powerful and celebrated mathematical reasoning which he had created. The explanation was the law of gravitation.

It occurred to Newton that if a diagram of the path of the Moon for any given period, say one minute, be made, it would be found that the Moon departs from a straight line during that period by a measurable distance. In other words, the Moon has been virtually pulled toward the Earth by an amount that is represented by the difference between its actual position at the end of the minute and the position it would occupy had it moved in a straight line, which according to Galileo's law of motion, it should follow unless some external force deflected it. By measuring the amount of deflection, he had a basis for determining the amount of the deflecting force. This deflection Newton found by his calculation to be thirteen feet. Obviously the force that acted on the Moon made it fall toward the Earth a distance of thirteen feet during the first minute of its fall.

Galileo had shown that the rapidity of a body's fall to the Earth increased at a uniform rate—what is now termed the acceleration of gravity. In other words, the higher the starting point of the fall, the greater will be the final velocity. Hence the amount of the attracting force is in some way related to the distance between the two bodies, a relation which Newton expressed by stating that the falling body is attracted to the Earth by a force which varies inversely as the square of the distance between them.

If the attracting force then varies inversely as the square of the distance, would the Moon drop toward the Earth at the calculated rate of thirteen feet in the first minute? That was the problem which presented itself to Newton. The mathematical solution was simple, based as it was on a comparison of the Moon's distance with the length of the Earth's radius. Unfortunately there were no accurate dimensions of the Earth available when Newton made his first calculation in 1666. Hence he found, on the basis of the erroneous data at his disposal, that the Moon fell toward the Earth fifteen instead of thirteen feet during the first minute, a discrepancy so great that he dismissed the matter from his mind.

When in 1682 his attention was called to a new and apparently accurate measurement of a degree of the Earth's meridian made by the French astronomer Picard, he attacked the problem anew. As he proceeded with his computation it became more and more certain that this time the result harmonized with the observed facts. So completely was he overwhelmed that he was forced to ask a friend to complete the simple calculation. When the computation was ended it was known that the force which causes bodies to fall to the Earth extends outward to the Moon, and that by reason of this force the Moon circles around the Earth.

It required but a slight stretch of the imagination to assume that a force which can span the distance between the Earth and the Moon may also span the distance from the Sun to the Earth and the other members of the solar system. That such is really the case, Newton proved by a mathematical calculation of the orbital motions of Jupiter's satellites and of the various planets.

These discoveries and fundamental principles enunciated by Newton were elaborated with great exactness in his 'Principia,' and the section which discusses the motions of the Moon, confessedly one of the most difficult problems in celestial mechanics, has been termed by Sir George Airy the greatest chapter on physical science ever written. That it has stood the test of time is demonstrated by the fact that Newton's results have scarcely been extended in the centuries which have elapsed since their publication. The entire work is a marvel of exact mathematical reasoning by "the greatest genius the world has ever produced," according to Lagrange's estimate of Newton's intellectual powers.

Galileo had experimentally shown before Newton that the rate at which two bodies fall to the ground from equal heights is independent of their weights. A mass of gold and a mass of lead, altho of unequal weight, reach the Earth at the same time if dropped simultaneously from the same height. Newton repeated the experiment very exactly. He realized as a result that weight (gravitation) is constant. But because a pound of lead weighs less than two pounds of lead (in other words, is attracted with one-half the force) merely for the reason that it contains less matter, he was forced to the conclusion that gravitation is dependent upon quantity of matter as well as distance. Thus he introduced the very difficult conception of mass as distinguished from weight, or the force of attraction exerted on it by the Earth. The former, of course, is absolute and constant, but the latter varies with the position of the material in question on the Earth's surface or elsewhere in the universe

If the mass of Venus is seven times that of Mars, then the force with which the Sun attracts Venus is seven times as great as that with which it would attract Mars if placed at the same distance; and therefore also the force with which Venus attracts the Sun is seven times as great as that with which Mars would attract the Sun if at an equal distance from it. Hence, in all cases of attraction, the force is proportional not only to the mass of the attracted body, but also to that of the attracting body as well as being inversely proportional to the square of the distance. Gravitation thus appears no longer as a property peculiar to the central body of a revolving system, but as belonging to a planet in just the same way as to the Sun, and to the Moon, or to a stone in just the same way as to the Earth.

Moreover, the fact that separate bodies on the surface of the Earth are attracted by the Earth and therefore in turn attract it, suggests that this power of attracting other bodies, which the celestial bodies are shown to possess, does not belong to each celestial body as a whole, but to the separate particles of which it is composed; so that, for example, the force with which Jupiter and the Sun attract each other is the result of compounding the forces with which the separate particles making up Jupiter attract the separate particles making up the Sun. Thus is suggested finally the law of gravitation in its most general form: 'Every particle of matter attracts every other particle with a force proportional to the mass of each and inversely proportional to the square of the distance between them.'

When Newton completed his 'Principia' astronomy became in the fullest sense an exact science. Given the positions, velocities and motions of the Sun, Earth, Moon and other planets, then the manner in which they interact on one another can be learned and even their form and dimensions determined. In short, astronomy, from a more or less mystical science became in earnest a mathematical science. When the motions and orbits of heavenly bodies were once observed, the positions of these bodies could be computed for future epochs.

In his 'Principia' Newton confines himself to the demonstration of the laws of gravitation. He says nothing about the means by which bodies are made to gravitate toward each other. His mind did not rest at this point. He felt that gravitation itself must be capable of being explained. It is known that he even suggested an explanation depending on the action of an ethereal medium pervading space. But with that wise moderation which is characteristic of all his investigations, he distinguished such speculations from what he had established by observation and demonstration and excluded from his 'Principia' all mention of the cause of gravitation, reserving his thoughts on this subject for the 'Queries' printed at the end of his 'Opticks.' The attempts which have been made since the time of Newton to solve this difficult question are few in number and have not yet led to any well-established result.

CHAPTER VI

To THE ancients as well as to the moderns the Sun and the Moon appeared not only the largest but the most important of all the celestial bodies. With the Sun and Moon five other conspicuous spheres eventually were linked, spheres distinguished by reason of their regular motions. These orbs, Mercury, Venus, Mars, Jupiter and Saturn, were named "planets" or "wanderers" to distinguish them from the "fixed" stars.

Venus, familiar as the evening star or the morning star, was discovered—it is claimed—by Pythagoras in the sixth century B.C., but even in the poems of Homer there are references to both stars without any indication of their identity. Jupiter, Venus, Mars and Saturn, ranking with the brightest of the stars, and Mercury, occasionally seen near the horizon just after sunset or before sunrise, all were known to the ancients. A study of their movements naturally led to the obvious conclusion that all these moving stars or planets were related in some way and that the motion of one was more or less dependent on the motions of the others. Hence it may be asserted that the ancient history of astronomy begins with the system of planets that revolve around the Sun.

What is the nature of these planets? Obviously they are not all alike in size or distance. Even to the naked eye their appearance seems to reveal conditions that need explanation. Early observation and study revealed the

57

fact that the planets occupied a section of the heavens where there were no so-called "fixed" stars. But later observation also revealed that, associated with the planets, are a number of smaller bodies of much the same nature known as "planetoids," or "asteroids," which, with a single exception, occupy the zone of the heavens between Mars and Jupiter. Lastly there are a large number of temporary visitors to this solar system known as the "comets." They plunge in from space, sweep around the sun and drift away by various paths or orbits, most of them never to return.

Planets, satellites, planetoids and comets comprise the Solar System. Vast and marvelously complete as that system is, it must be admitted that it is but a part of the great universe. It may be, as there is some reason to suppose, that this Solar System is but one of many similar systems scattered throughout the universe and that each of these—including that in which the Earth is situate—is in turn wheeling about some central orb inexpressibly distant. The Solar System to which the Earth belongs is merely a type and not a unique example of planetary order.

The intellectual rise in Astronomy is nowhere more clearly revealed than in the history of man's conception of the Solar System. Perhaps the first inquiry that must have flashed across the mind of a thinking Chaldean or Greek concerned itself with the distances of the heavenly bodies. How far away are the planets? How is their distance measured? The second question concerned itself with their motions, Whither do they drift and why? Around these questions cluster a group of vague guesses, fruitless speculations and poetic fancies, from which at last a scientific method was evolved for measuring planetary distances and accounting for planetary movements. It was not until comparatively late in astronomical history that means were devised for ascertaining the physical condition of each planet.

The distances of the planets, small as they seem in com-

parison with sidereal measurements, are felt to be immense.
Using only round numbers, which are sufficiently accurate
for the present purpose, the planet Neptune, the outermost
known member of our system, is 2,800,000,000 miles from
the Sun. In a cord twenty-eight feet long each single foot
will represent a hundred million miles. On such a scale
a map of the United States could not be seen without the
aid of a microscope. Suppose a bead were placed at each
end of this line, one representing the Sun, the other Neptune. Between the two, other beads will represent the other
planets. One nearly four inches from that representing
the Sun will be Mercury; another, at about seven inches,
Venus; a third, at eleven inches, the Earth; a fourth, at

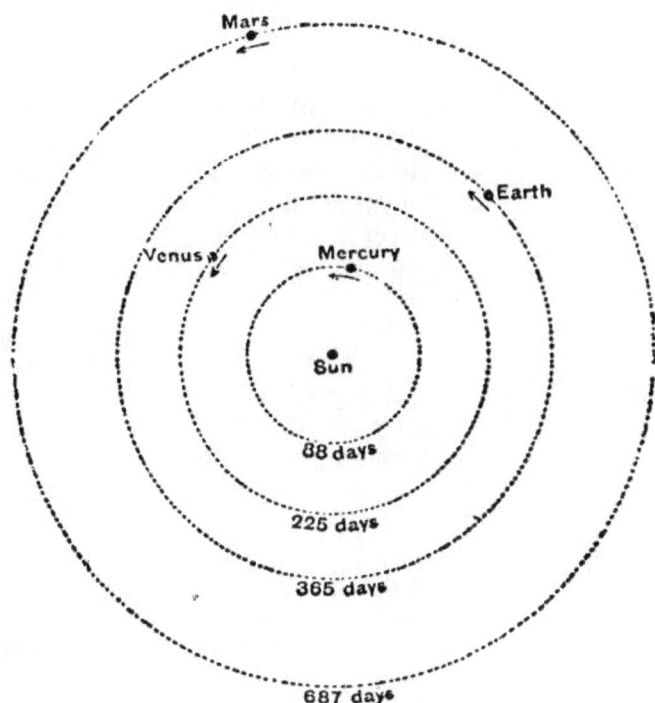

Fig. 7 —THE ORBITS OF THE FOUR INTERIOR PLANETS.

seventeen inches, Mars; a fifth, at about five feet, Jupiter; a sixth, at nine feet, Saturn; a seventh, at eighteen feet, Uranus, and an eighth, Neptune, at the end.

The mean distances of the planets from the Sun are as follows:

	MILES.
Mercury	36,000,000
Venus	67,200,000
Earth	92,900,000
Mars	141,500,000
Jupiter	483,300,000
Saturn	886,000,000
Uranus	1,781,900,000
Neptune	2,791,600,000

Attempts to measure some of these distances approximately are found in early times. The idea that some of the planets must be nearer the Earth than others must have been suggested by eclipses and occultations—*i.e.*, passage of the Moon over the Sun and over a planet or fixed star. No direct means being available for determining the distance, rapidity of motion anciently was employed as a test of probable nearness. The stars being seen above, it was but natural to think of the most distant celestial bodies as the highest, and accordingly Saturn, Jupiter and Mars, being beyond the Sun, were called "superior planets" as distinguished from the two "inferior planets," Venus and Mercury. Uranus and Neptune are modern additions to the solar system and could not have been included in the hypothesis. Aristotle (384-322 B.C.), for example, arrived at the conclusion that the planets are farther off than the Sun and Moon as the result of an occultation of Mars by the Moon and as the result of similar observations made in the case of other planets by the Egyptians and Babylonians. Ptolemy (second century A.D.), altho far more original and daring in his astronomical conceptions than Aristotle, was able to add but little toward a solution of the problem. He expressly states that he had no means of estimating numerically the distances of the planets or even of knowing the order of the distance of the several planets. He followed tradition in conjecturally accepting rapidity of motion as a test of

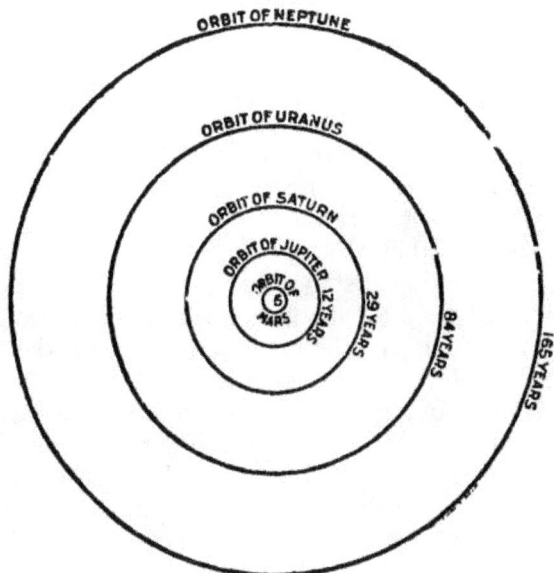

Fig. 8 —THE ORBITS OF THE FIVE EXTERIOR PLANETS.

nearness and placed Mars, Jupiter, Saturn (which perform the circuit of the celestial sphere in about 2, 12 and 29 years, respectively) beyond the Sun in that order. As Venus and Mercury accompany the Sun, and may therefore be regarded as on the average performing their revolutions in a year, the test to some extent failed in their case, but Ptolemy again accepted the opinion of the "ancient mathematicians"—probably the Chaldeans—that Mercury and Venus lie between the Sun and Moon, Mercury being the nearer to the Earth.

Copernicus gave the first glimpse of the truth. To quote Berry in his "Short History of Astronomy": "From the fact that Venus and Mercury were never seen very far from the Sun, it could be inferred that their paths were nearer to the Sun than that of the Earth, Mercury being the nearer to the Sun of the two, because never seen so far from it in the sky as Venus. The other three planets, being seen at times in a direction opposite to that of the Sun, must necessarily revolve round the Sun in orbits larger than that of the Earth, a view confirmed by the fact that they were brightest when opposite the Sun (in which positions they would be nearest to us). The order of their respective distances from the Sun could be at once inferred from the disturbing effects produced on their apparent motions by the motion of the Earth. Saturn being least affected, must on the whole be farthest from the Earth, Jupiter next and Mars next. The Earth thus became one of six planets revolving round the Sun, the order of distance—Mercury, Venus, Earth, Mars, Jupiter, Saturn—being also in accordance with the rates of motion round the Sun, Mercury performing its revolution most rapidly (in about 88 days), Saturn most slowly (in about 30 years)."

It was not until John Kepler (1571-1630) published his "Epitome of the Copernican Astronomy," his "Harmony of the World" and a treatise on "Comets" that astronomers were given a definite formula which enabled them to determine planetary distances with any exactitude. Kepler's speculative and mystic temperament led him constantly to search for relations between the various numerical quantities occurring in the Solar System. By a happy inspiration he tried to discover a relation between the sizes of the orbits of the various planets and their times of revolution round the Sun. After a number of unsuccessful attempts he discovered a simple and important relation commonly known as Kepler's third law:

"The squares of the times of revolution of any two planets (including the Earth) about the Sun are proportional to the cubes of their mean distances from the Sun."

In other words, given the periods, there is need only to find the interval between any two of them in order to infer at once the distance separating them all from one another and from the Sun. Here was the plan. What was next to be discovered was the scale upon which the plan was to be drawn. There must be first a trustworthy measure of the distance of a single planet from the Sun, the Earth, for example, and the problem would be solved.

How is this measure to be obtained? Sir Robert Ball in his "Story of the Heavens," gives this simple example for partial explanation: "Stand near a window where you can look at buildings . . . or at any distant object. Place on the glass a thin strip of paper vertically in the middle of one of the panes. Close the right eye and note with the left eye the position of the strip of paper relatively to the objects in the background. Then, while still remaining in the same position, close the left eye and again observe the position of the strip of paper with the right eye. You will find that the position of the paper on the background has changed.

"Move closer to the window and repeat the observation, and you find that the apparent displacement of the strip increases. Move away from the window and the displacement decreases. Move to the other side of the room, the displacement is much less, tho probably still visible. We thus see that the change in the apparent place of the strip of paper, as viewed with the right eye or the left eye, varies in amount as the distance changes; but it varies in the opposite way to the distance, for as either becomes greater the other becomes less. We can thus associate with each particular distance a corresponding particular displacement. From this it will be easy to infer that, if we have the means of measuring the amount of displacement, then we have the means of calculating the distance from the observer to the window. It is this principle applied on a gigantic scale which enables us to measure the distances of the heavenly bodies.

"Look, for instance, at the planet Venus; let this correspond to the strip of paper and let the Sun, on which Venus is seen in the act of transit, be the background. Instead of the two eyes of the observer, we now place two observatories in distant regions of the Earth; we look at Venus from one observatory, we also look at it from the other; we measure the amount of displacement and from that we calculate the distance of the planet. All depends, then, on the means which we have of measuring the displacement of Venus as viewed from the two different stations."

Two observers standing upon the Earth must be some thousands of miles apart in order to see the position of the Moon altered with regard to the starry background to obtain the necessary data upon which to ground their calculations. The change of position thus offered by one side of the Earth's surface at a time is not sufficient, however, to displace any but the nearest celestial bodies. When there is occasion to go farther afield, a greater change of place must be sought. This can be obtained as a consequence of the Earth's movement around the Sun. Observations, taken several days apart, will show the effect of the Earth's change of place during the interval upon the positions of the other bodies of our system. But when the depths of space beyond are to be sounded and an effort is made to reach out for the purpose of measuring the distance of the nearest star the utmost change of place is necessitated. This results from the long journey of many millions of miles which the Earth performs around the Sun during the course of each year. Still, even this last change of place, great as it seems in comparison with terrestrial measurements, is insufficient to show anything more than the tiniest displacements in a paltry forty-three out of the entire host of stars.

It is thus readily realized with what an enormous disadvantage the ancients coped. The measuring instruments at their command were utterly inadequate to detect such small displacements. It was reserved for the telescope to reveal them, and even then it required the great telescopes of recent times to show the slight changes in the position of the nearer stars which were caused by the Earth's being at one time at one end of its orbit and some six months later at the other end—stations separated by a gulf of about 186,000,000 miles.

It was from an opposition of Mars observed in 1672 by John Richer (?-1696) at Cayenne in concert with Giovanni Domenico Cassini at Paris that the first scientific estimate of the Sun's distance was derived. The Sun appeared to be nearly 87,000,000 miles away. John Flamsteed (1646-1720), the first Astronomer Royal of England, deduced 81,700,000 from his independent observations of the same occurrence. Jean Picard's (1620-1682) later result was just one-half Flamsteed's (41,000,000). Philippe De Lahire thought that the Earth must be separated from the Sun by at least 136,000,000 miles. The transits of Venus in 1761 and 1769 were employed, after other attempts had been made, to measure the Sun's distance.

The transit of 1769 is of particular interest, not only for a fairly good determination of the Sun's distance, but also for the reason that the celebrated Captain Cook was commissioned to sail to Otaheite for the purpose of witnessing the transit of Venus. At Otaheite, on June 3d, the phenomenon was carefully observed and measured. Simultaneously with these observations others were obtained in Europe and elsewhere. From a combination of all the observations, an approximate knowledge of the Sun's distance was gained. The most complete discussion of these observations did not, however, take place for some time. It was not until the year 1824 that the illustrious Johann Franz Encke computed the distance of the Sun and gave as the definite result 95,000,000 miles. Later Urbain Jean Joseph Le Verrier (1870) reduced the estimate to 91,320,000 miles, which held good until Prof. Simon Newcomb in 1882 gave the figure 92,475,000 miles. In 1900 nearly all the observatories of the world under the direction of Maurice Loewy and the French Academy of Science began a new computation which will lead to more exact results. The old problem of measuring a planet's distance from the Sun its not yet completely solved. If Sir David Gill's plan of basing a new set of calculations on the opposition of Eros in 1931 is carried into execution, the Sun's distance will be ascertained to within 10,000 miles. Present knowledge declares the distance of the planets from the Sun with an error not exceeding one-fiftieth of one per cent.

CHAPTER VII

THE motions of the planets also formed the basis for archaic theorizing. That the planets move, the ancients were fully aware, for the very word "planet" means "wanderer." The strip of the celestial sphere through which move the Sun, the Moon and the five planets known to the ancients (Mercury, Venus, Mars, Jupiter and Saturn) was called the Zodiac, because the constellations in it were named after living things, with one exception. The Zodiac was divided into twelve equal parts, the "signs of the Zodiac," through one of which the Sun passed every month, so that its position could be roughly given by stating in what sign it was. The stars in each sign were formed into a constellation, the sign and the constellation each receiving the same name. The relative movements of the planets as the ancients conceived them are thus summarized by Berry: "In Pythagoras occurs perhaps for the first time an idea which had an extremely important influence on ancient and medieval astronomy. Not only were the stars supposed to be attached to a crystal sphere, which revolved daily on an axis through the Earth, but each of the seven planets (the Sun and Moon being included) moved on a sphere of its own. The distances of these spheres from the Earth were fixed in accordance with certain speculative notions of Pythagoras as to numbers and music; hence the spheres as they revolved produced harmonious sounds which specially gifted persons might at times hear. This is the origin of the idea of the music of the spheres which recurs continually in medieval speculation and is found occasionally in modern literature. At a later stage these spheres of Pythagoras were developed into a scientific representation of the motions of the celestial bodies, which remained the basis of astronomy till the time of Kepler."

Philolaus, the Pythagorean, who lived about a century after his master, introduced for the first time the idea of the motion of the Earth. He appears to have regarded the Earth, as well as the Sun, Moon and five planets, as revolving round some central fire, the Earth rotating on its own axis as it revolved, apparently in order to insure that the central fire should always remain invisible to the inhabitants of the known part of the Earth. Altho pure fancy, the idea of Philolaus was a valuable contribution to astronomical thought.

Despite the immense influence of the Pythagoreans, most Greeks shared Plato's idea that any careful study of celestial motions was degrading rather than elevating, for the whole subject smacked too much of the unesthetic section of geometry. Still, Plato (429-347 B.C.) did give a short account of the celestial bodies, according to which the Sun, Moon, planets and fixed stars revolve on eight concentric and closely fitting wheels or circles around an axis passing through the Earth.

This idea of Plato's was more or less followed by later philosophers. Thus Eudoxus of Cnidus (409-356 B.C.) attempted to explain the more obvious peculiarities of planetary motion by means of a combination of uniform circular motions. The celestial motions were to some extent explained by means of a system of 27 spheres, 1 for the stars, 6 for the Sun and Moon, 20 for the planets. There is no clear evidence that Eudoxus made any serious attempt to arrange either the size or the time of revolution of the spheres so as to produce a precise agreement with the observed motion of the celestial bodies, tho he knew

with considerable accuracy the time required by each
planet to return to the same position with respect to the
Sun; in other words, his scheme represented the celestial
motions qualitatively but not quantitatively.

Aristotle adopted this scheme of Eudoxus, but need-
lessly complicated it by treating the spheres as material
bodies and added 22 more spheres, thus making 56 in all.
He argued against the possibility of the Earth's revolving
around the Sun on the ground that there ought to be a
corresponding apparent motion of the stars, an objection
finally disposed of only during the nineteenth century,
when it was discovered that this motion can be seen only
in a few cases because of the unutterably great distance of
the stars.

No substantial advance can be noted until Hipparchus
(160-125 B.C.) made an extensive series of observations
with all the accuracy that his instruments would permit
and critically made use of old observations for comparison
with later ones so as to discover astronomical changes too
slow to be detected in a single lifetime—an essentially
modern method. He systematically employed a geometrical scheme (that of eccentrics and epicycles) for the representations of the motions of the Sun and the Moon, a
mode suggested in substance by Apollonius of Perga, who
flourished in the third century B.C.

The great services rendered to astronomy by Hipparchus
can hardly be better expressed than in the words of the
great French historian of astronomy, Delambre, who is in
general no lenient critic: "When we consider all that Hipparchus invented or perfected and reflect upon the number
of his works and the mass of calculations which they
imply, we must regard him as one of the most astonishing
men of antiquity and as the greatest of all in the sciences
which are not purely speculative and which require a combination of geometrical knowledge with a knowledge of
phenomena to be observed only by diligent attention and
refined instruments."

The last great name encountered in tracing the record of changing conceptions of planetary motions is that of Ptolemy (100-170 A.D.), whose reputation rests on his "Almagest," which may be regarded as the astronomical gospel of the Middle Ages. Hipparchus, as we have seen, found the current representations of the planetary motions inaccurate and collected a number of new observations. These, with fresh observations of his own, Ptolemy employed in order to construct an improved planetary system. Following the idea of Hipparchus, Ptolemy thought that the Sun and Moon moved in circular orbits around the Earth as a center. Ptolemy's chief work was to expand the system of epicycles so that it could explain discrepancies between theory and observation, discrepancies overlooked or ignored by Hipparchus. The deviations of the planets from the ecliptic, for example, were accounted for by tilting up the planes of the epicycles. Thus with the aid of the system of Hipparchus, supplemented with his own idea of tilting epicycles, he worked out with great care and labor the motions of the planets. Altho the Hipparchian-Ptolemaic doctrine was framed on an extravagant estimate of the importance of the Earth in the scheme of the heavens, yet it must be admitted that the apparent movements of the celestial bodies were thus accounted for with considerable accuracy. For fourteen centuries the Almagest was regarded as the final authority on all questions of astronomy and it may be considered as the loftiest piece of calculation appertaining to the Ancient World.

CHAPTER VIII

THE Ptolemaic system of astronomy was discredited only at an epoch nearly simultaneous with that of the discovery of the New World by Columbus. The true arrangement of the solar system was then expounded by Nicholas Copernicus (1473-1543) in the great work, "De Revolutionibus," to which he devoted his life. The first principle established by these labors showed the diurnal movement of the heavens to be due to the rotation of the Earth on its axis. Copernicus pointed out the fundamental difference between real motions and apparent motions. He proved that the appearances presented in the daily rising and setting of the Sun and the stars could be accounted for by the supposition that the Earth rotated, just as satisfactorily as by the more cumbrous supposition of Hipparchus and Ptolemy. He showed, moreover, that if the ancient supposition were true, the stars must have an almost infinite velocity and declared that the rotation of the entire universe around the Earth was clearly preposterous.

The second great principle, which has conferred immortal glory upon Copernicus, assigned to the Earth its true position in the universe. Copernicus transferred the center, about which all the planets revolve, from the Earth to the Sun, and he established the somewhat crushing truth that the Earth is merely a planet, pursuing a track between the paths of Venus and of Mars and subordinated like all the other planets to the supreme sway of the Sun. This great revolution swept from Astronomy those distorted views of the Earth's importance which arose, perhaps not unnaturally, from the fact that the observers chanced to live on this particular planet. Whether the actual services rendered by Copernicus are commensurate with his fame may be doubted. He labored under the weight of an ecclesiastical tradition that could not be abandoned without some risk. He was a bold man indeed who dared to overthrow or even to question orthodoxy and to diminish the Earth's overshadowing importance in the Solar System.

The Copernican system was not flawless either in theory or logic and many objections could be made to it, particularly by an astronomer who had observed and studied the movements of the heavenly bodies. After the example of the ancients, Copernicus assumed as an axiom the uniform, circular motion of the planets, and, as the only motions which are observed are in a state of incessant variation, he was obliged, in order to explain the inequalities to suppose a different center for each of the orbits. The Sun was placed within the orbit of each of the planets, but not in the center of any of them. In other words, he still adhered to a system of epicycles. Consequently the Sun performed no other office than to distribute light and heat. Excluded from any influence on the system, the Sun became a stranger to all the motions. The "fixed" stars were alleged to be stationary, and it was necessary to suppose that they were almost infinitely distant, inasmuch as they always seemed to preserve the same position when viewed from the opposite sides of the Earth's orbit.

While various astronomers showed some disposition to accept the Copernican teaching, most of them were bitterly opposed to it on ecclesiastical, traditionary and scientific grounds. Tycho Brahe (1546-1601) was the most distinguished of these opponents. Being an indefatigable observer and practically the first to realize the value of continuous observation, he enriched astronomy by a star catalogue and studies of the movements of the other heavenly bodies. Tycho accepted the Copernican conception of a central Sun, but rejected the idea that the Earth

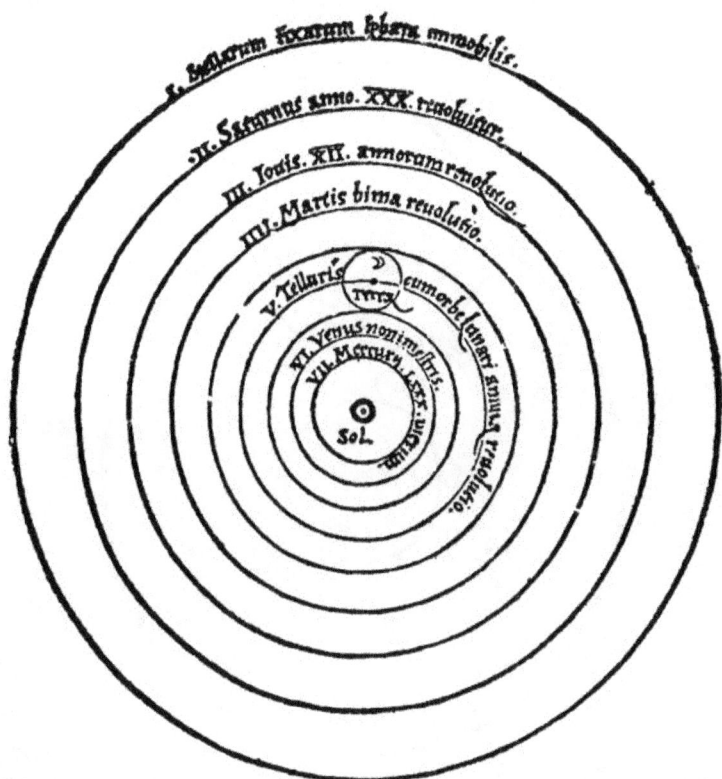

Fig. 9 —THE SOLAR SYSTEM ACCORDING TO COPERNICUS. (From the De Revolutionibus.)

moved. Thus he sought to effect a compromise between the Ptolemaic and Copernican systems. It was the study of a comet in 1577 that led Tycho to formulate his ideas of the solar system. He believed that the comet (X) as shown in the accompanying diagram was revolving around the Sun at a distance greater than that of Venus and assumed that both the Sun (C) and the Earth (A) were centers of revolving systems, the five planets revolving around the Sun and the entire system in turn moving around the Earth. This incorrect proposition, which

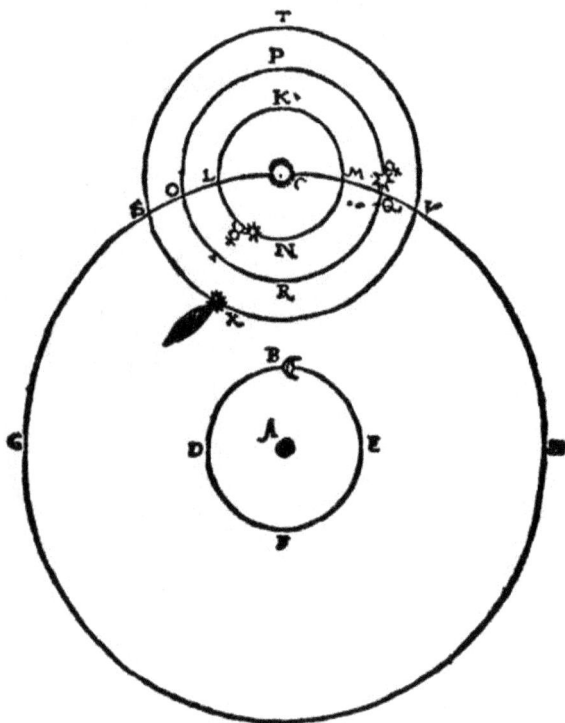

Fig. 10 —Tycho's System of the World. (From his book on the Comet of 1577.)

was one of the least of Tycho Brahe's contributions to astronomical science, is significant, showing as it does how difficult it was for the principles of Copernicus firmly to establish themselves and planetary motion to be explained satisfactorily.

Whatever Tycho may have thought of the Copernican

system, his contemporary, Galileo (1564-1642), was willing to accept it. It has been shown how Galileo with the telescope of his invention was able to extend astronomical science and to introduce new methods of observation, which came naturally to one who was a leader in the experimental science of his time. But even before his work with the telescope Galileo had adopted the astronomical views of Copernicus and collected arguments for their support. He was able in 1604 to confirm the discovery of Tycho Brahe that changes take place in the heavens beyond the planets and that there was an important region beyond the Earth and its immediate surroundings. As was but natural the use of the telescope broadened Galileo's horizon, and, true scientist that he was, he immediately brought to bear his new discoveries on the fundamental conceptions.

Thus his discovery of the satellites revolving around Jupiter as the planets themselves revolved around the Sun not only rendered necessary the explanation of these new bodies, but dealt a serious blow at the infallibility of Aristotle and Ptolemy, neither of whom had any idea of the existence of these satellites. Further support was given to the Copernican theory by the ocular demonstration of these satellites revolving around Jupiter and not dropping behind, just as the Moon was required to move around the Earth, a mechanical difficulty brought forward by the opponents of the Copernican idea. As Galileo developed his astronomical ideas and discoveries he naturally came into conflict with ecclesiastical authority and there began the unfortunate controversy as to the relative validity in scientific matters of observation and reasoning on the one hand and the authority of the Church and Bible on the other.

Controversies such as this were conspicuous in the latter part of Galileo's life. They culminated in his famous trial and formal abjuration of his alleged errors and in his conviction "of believing and holding the doctrines

—false and contrary to the Holy and Divine Scriptures— that the Sun is the center of the world and that it does not move from east to west, and that the Earth does move and is not the center of the world; also that an opinion can be held and supported as probable after it has been declared and decreed contrary to the Holy Scriptures." Despite Galileo's abjuration, his general attitude toward the Church and the Bible is contained in his approval of the saying of Cardinal Baronius, "That the intention of the Holy Ghost is to teach us not how the heavens go, but how to go to heaven." His attempts to explain and demonstrate the Copernican system in his great astronomical treatise, "Dialog on the Two Chief Systems of the World, the Ptolemaic and Copernican," led to his trial and conviction before the Inquisition.

Kepler, another of Galileo's contemporaries, did more even than the great Italian to bring about a proper conception of the solar system and the motions of the planets. A student under Tycho, it was but natural that Kepler should have imbibed from his master a respect for systematic observation, regardless of the correctness or incorrectness of Copernican views. As a result Kepler early adopted the Copernican doctrine, opposed tho it was by his master. His observations led him to the conclusion, however, that even Copernicus had not revealed all the mysteries of planetary motion and that the hypothetical circles in which the planets revolved around the Sun, according to Copernicus, did not agree with the paths observed. Under the instruction of Tycho, Kepler addressed himself to the problems involved in the planet Mars, whose positions as seen in the sky were a combined result of its own motion and that of the Earth, as both move around the Sun. Actual observation of the planet and the consideration of various geometrical theories that suggested themselves eventually led to the conception that the path of the planet must be some form of an oval.

Finally Kepler reached the conclusion that instead of being circular, the planet's motion must lie in the simple curve known as an ellipse and formed by taking an oblique section of a cone. While the circle has but a single center, the ellipse depends for its form upon two fixed points, each of which is termed a focus. It can be drawn by

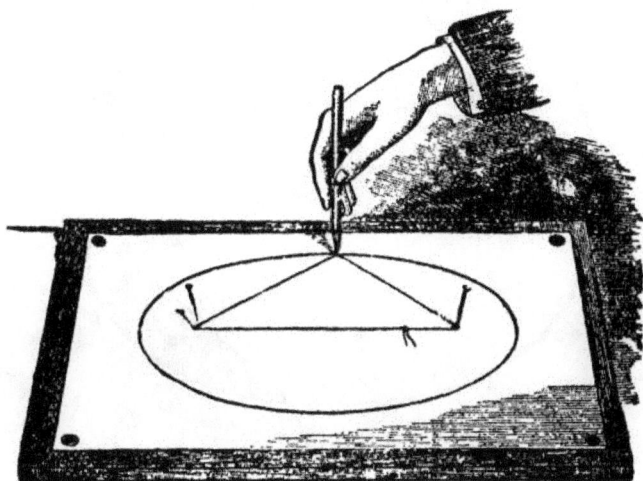

Fig. 11 —DRAWING AN ELLIPSE

using two pins stuck in a sheet of paper and by inserting a pencil within a loop of string that also includes the two pins. The curve may be traced by moving the pencil, while the string is kept taut. It will be found that if the two points are kept close together the curve approaches in form a circle, while if they are separated the figure becomes elongated and possesses what the mathematicians term greater eccentricity. At any rate, every point on the curve is such that the sum of its distance from the two foci is always the same. Kepler found that the Sun was at one focus. When the planet was near that focus, it moved with greater velocity than when at the opposite part of its orbit. The speed of motion, however, was always proportional to the areas swept out by a straight line from the Sun in equal intervals of time. In other words, there were formulated the now famous first and second laws of Kepler as follows:

1. The planet describes an ellipse, the Sun being in one focus.

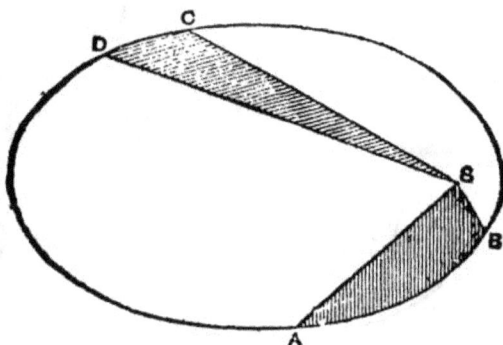

Fig. 12 —EQUAL AREAS IN EQUAL TIMES.

2. The straight line joining the planet to the Sun sweeps out equal areas in any two equal intervals of time.

Kepler not only established these laws for Mars, but immediately applied his principle to the Earth and then claimed (without proof, however) in his "Epitome of the Copernican Astronomy" that these two fundamental laws applied also to all the planets and to the motions of the Moon. Accompanying these two laws was the third, already discussed, in which it is stated that the squares of the times of revolution of any two planets (including the Earth) about the Sun are proportional to the cubes of their mean distances from the Sun.

It was the disclosure of these wonderfully simple relations that laid the foundation for the Newtonian law of gravitation. Contemporary judgment, of course, could not anticipate the culmination of a later generation. What it could understand was that the first law of Kepler attacked one of the most time-honored or metaphysical conceptions —the Aristotelian idea that the circle is the perfect figure and that planetary motions consequently must be circular. Not even Copernicus had doubted the validity of this assumption.

Kepler was too great a genius to rest content with the mere observation that the planets move in ellipses. Next he desired to determine why they do so move. It remained for Isaac Newton (1643-1727) to answer the question satisfactorily; yet Kepler had a curious premonition of the law of gravitation. "Whereas the Ptolemaic system," comments Berry, "required a number of motions round mere geometrical points, centers of epicycles or eccentrics, equants, etc., unoccupied by any real body, and many such motions were still required by Copernicus, Kepler's scheme of the solar system placed a real body, the Sun, at the most important point connected with the path of each planet and dealt similarly with the Moon's motion round the Earth and with that of the four satellites round Jupiter. Motions of revolution came in fact to be associated not with some 'central point' but with some 'central body,' and it became therefore an inquiry of interest to ascertain if there were any connection between the motion and the central body. The property possessed by a magnet of attracting a piece of iron at some little distance from it suggested a possible analogy to Kepler, who had read with care and was evidently impressed by the treatise 'On the Magnet' ('De Magnete'), published in 1600 by William Gilbert of Colchester (1540-1603). He suggested that the planets might thus be regarded as sharing to some extent the Sun's own motion of revolution. In other words, a certain 'carrying virtue' spread out from the Sun, with or like the rays of light and heat, and tried to carry the planets round with the Sun." Kepler says himself in his "Epitome":

"There is, therefore, a conflict between the carrying power of the Sun and the impotence or material sluggishness (inertia) of the planet; each enjoys some measure of victory, for the former moves the planet from its position and the latter frees the planet's body to some extent from the bonds in which it is thus held, . . . but only to be captured again by another portion of this rotatory virtue."

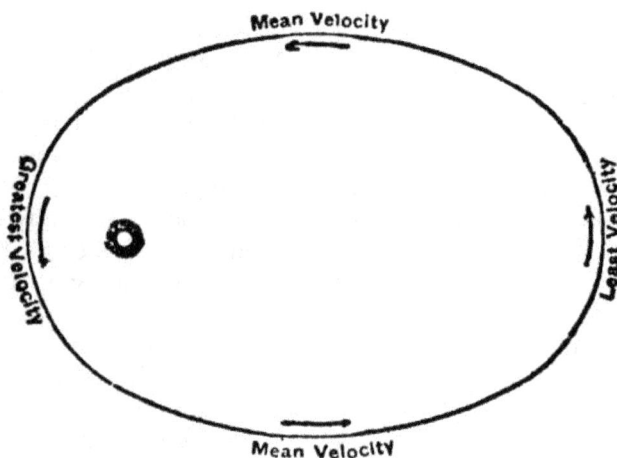

Fig. 13 —Varying Velocity of Elliptic Motion.

Thus is faintly indicated the great theory of gravitation which, as developed by Newton, was to supply a satisfactory explanation of planetary motion and which is the underlying basis of all modern astronomy.

Newton had become convinced that the attracting power of the Earth extended even to the Moon, and that the acceleration produced in any body—whether it be as distant as the Moon or close to the Earth—was inversely proportional to the square of the distance from the Earth's center and also proportional to the mass of the body.

Then he found that the motions of the planets could be explained by an attraction toward the Sun, which produced an acceleration inversely proportional to the square of the distance from the center of the Sun, not only in the same planet in different parts of its path, but also in different planets.

Again it follows from this that the Sun attracts any planet with a force inversely proportional to the square of the distance of the planet from the Sun's center and also proportional to the mass of the planet. Accordingly, if the Earth or Sun attracts a body, the body must exert a similar force on the Earth or the Sun, and gravitation is not only a property of the central body of a revolving system, but belongs to every planet in just the same way as to the Sun and to a Moon or to a stone just as to the Earth.

After Newton had established provisionally the law of gravitation and the laws of motion, it remained for him to prove that the observed motions of the planets agreed with those calculated. A situation of the greatest complexity, however, was relieved by the fact that the mass of even the largest planets is so very much less than that of the Sun that the motion of any single planet is affected but slightly by the others, and it may be assumed to be moving very nearly as it would if the other planets did not exist, due allowance being made subsequently for minor disturbances or perturbations produced in its path. One by one the various irregularities observed were explained, and the motion of the Moon and its various eccentricities were computed with accurate numerical results.

For many years the solar system remained in its fancied integrity. There was no change in the original five planets and the number of satellites first discovered by Galileo and added to by subsequent observers had reached an apparent culmination when G. D. Cassini (1625-1712) had detected the second pair of satellites of Saturn in 1684. Accordingly, when Herschel, following his custom of making a review of the heavens with each new telescope that he constructed, found March 13, 1781, with a Newtonian telescope, seven feet in length, a small ·star which appeared so much larger than its companions and of such uncommon appearance, he suspected it to be a comet. Further study revealed that it was more than a comet and of far greater interest. When heedfully observed and its path calculated, it was found that no ordinary cometary orbit would in any way fit its motion.

Anders Johann Lexell (1740-1784) first recognised that Herschel's body was not a comet but a new planet revolving around the Sun in a nearly circular path and at a distance about nineteen times that of the Earth and nearly double that of Saturn. A vain attempt by Herschel to name the new planet after his royal patron, George III., "Georgium Sidus," finally resulted in the proposal and acceptance by British and continental astronomers alike of the name Uranus, which harmonized with the names of the other planets.

This discovery was of especial interest, inasmuch as Johann Daniel Titius (?-1796), a professor at Wittenberg, had pointed out the remarkable symmetry in the disposition of the planets. In a note, published in 1772, he showed that the distance of the six known planets from the Sun could be represented with a close approach to accuracy by a certain series of numbers increasing in the regular progression, 0, 3, 6, 12, 24, 48, etc. Adding 4 to each number, the results would give the relative distances of the six known planets from the Sun. In applying this law (which does not hold good in the case of Neptune) it was found that the term of the series following that which corresponded with the orbit of Mars was not represented in the list of planets. Accordingly Johann Elbert Bode (1747-1826), a German astronomer, assumed a hypothetical planet to take this place. When Uranus was discovered its distance was found to fall but slightly short of perfect conformity with the law of Titius, and it stimu-

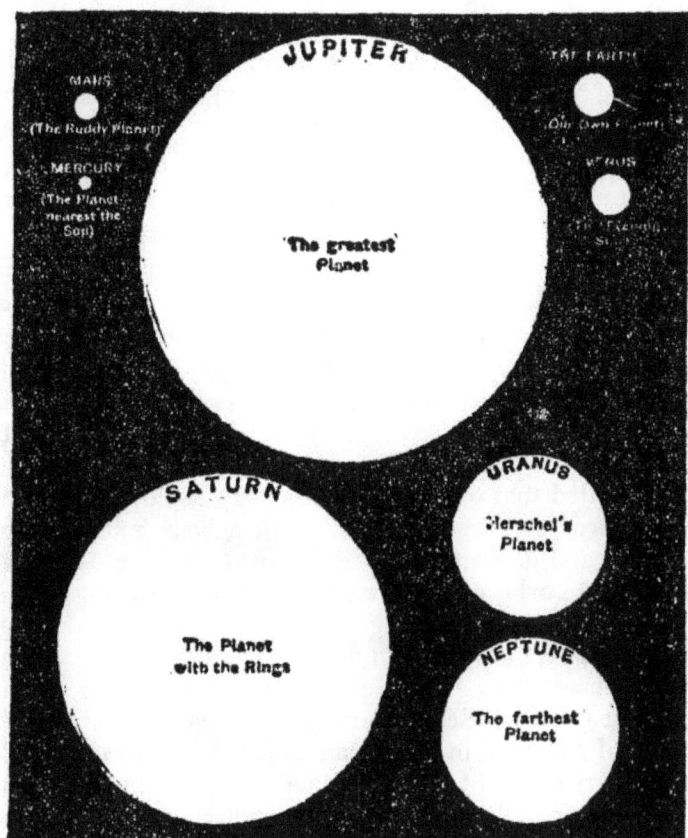

MARS
(The Ruddy Planet)

MERCURY
(The Planet nearest the Sun)

JUPITER

The greatest Planet

THE EARTH
(Our Own Planet)

VENUS
(The Evening Star)

SATURN

URANUS

Herschel's Planet

The Planet with the Rings

NEPTUNE

The farthest Planet

Fig. 14 —COMPARATIVE SIZES OF THE PLANETS.

lated the search for a new body, which, as will be seen in the chapter on the planetoids, proved to be the small planet Ceres.

The study of Uranus after its discovery by Herschel furnished many difficulties to astronomers. Despite the most careful calculations of the movements of the planet through more than a century's observations, the conclusion was reached that considerable errors existed or that the planet was wandering from its course. In fact, these disturbances had aroused the interest of several mathematicians and astronomers, and a young English student, John Couch Adams (1819-1892), soon after his graduation from Cambridge, communicated in 1845 to the Astronomer Royal numerical estimates of the elements and orbits of the unknown planet which he assumed was acting on Uranus and was the cause of the disturbances. In fact, he indicated the actual place in the heavens of the hypothetical planet. At practically the same time a French astronomer, Urbain Leverrier (1811-1877), who had made a careful study of the solar system in response to a request from Dominique F. J. Arago, the head of the French Observatory, prepared an elaborate memoir in which he demonstrated that only an exterior body could produce the disturbances observed and that such a body must occupy a certain and determinate position in the zodiac. He also assigned the orbit of the disturbing body, indicating that it would be as visible and bright as a star of the eighth magnitude. In fact, he supplied data to Professor Galle, of the Berlin Observatory, which enabled that astronomer to find in the heavens on September 23, 1846, within less than a degree of the spot indicated an object with a measurable disk. A reasonably complete map of this portion of the sky, in which all the stars were noted, proved beyond question that the object was not a star, while its movement as predicted was ample confirmation of its planetary nature. Adams' work, which antedated that of Leverrier, had not received attention originally in Great Britain at the hands of the Astronomer Royal, but as the matter assumed importance the Cambridge Observatory also participated in the search and on September 29, 1846, the planet was seen again. Thus Neptune was discovered. To show the rapidity of astronomical research in the nineteenth century it may be remarked that it required but seventeen days for the discovery of a satellite by Lassell (1799-1880) with a two-foot reflecting telescope.

Astronomers have suspected the existence of still other planets, and the belief has been expressed that such a body exists nearer to the Sun than Mercury, which, as has been seen, enjoys the reputation of being the closest of all the planets to the central luminary. The average distance of Mercury from the Sun is about 36,000,000 miles,

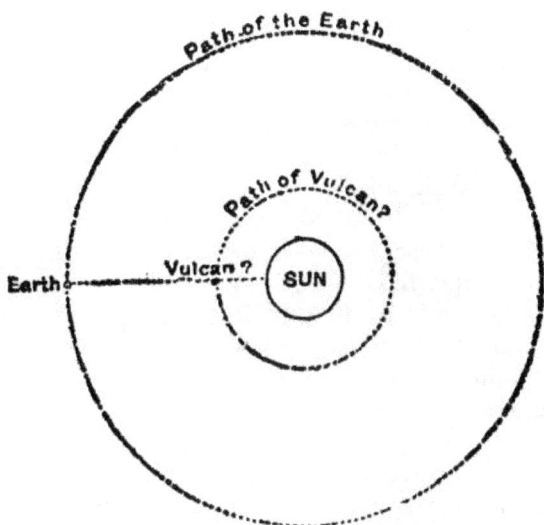

Fig. 15 —The Transit of "Vulcan," the Supposed Intramercurial Planet.

so that there would be space enough for such a planet. Its peculiar position in close proximity to the Sun, however, would act against its being observed. A small luminous point in this position would be altogether invisible, even with the best modern telescope, while its setting and rising, simultaneous almost with those of the Sun, make it invisible at these times even under the most favorable conditions. If this planet should pass across the Sun's disk just as do Mercury and Venus, it would be seen. While from its size it would be much less of a spectacle than the two planets mentioned, it might be detected. Claims have been advanced by astronomers that they have seen such a transit of a small spot.

The first suggestion of an intra-Mercurial planet came from the distinguished French astronomer, Leverrier, who in 1859 advanced such a hypothesis in an attempt to explain the movements of the planet Mercury. His theory involved a body of about the size of Mercury revolving at somewhat less than half its mean distance from the Sun, or at a greater distance if of less mass and vice versa, whose motion in great part would explain the irregularities observed. In the same year Dr. Lescarbault at Orgères maintained that he had observed such a body crossing the Sun's disk, and the name of Vulcan was bestowed upon it. Several astronomers claimed to have seen the new planet. Their observations were not well authenticated, and on the dates fixed for the probable transits no trace of the planet could be found. The strongest test was the examination of the sky at the time of a solar eclipse, for then the light of the Sun was cut off and a strange body could be readily identified. Despite a careful watch at subsequent eclipses and an examination of photographic plates, only negative results have been obtained. To-day the belief that there is any body of considerable size within the orbit of Mercury is held only by a few astronomers and very guardedly stated.

If the search for an intra-Mercurial planet was unsuccessful, it has in no way deterred astronomers from endeavoring to find other unrecognised members of the solar system. Much interest has been aroused by a hypothetical ultra-Neptunian planet, which of course would be the furthest from the Sun of all the members of the solar system. The basis for such a hypothesis is the reduction of observations made of the positions and motions of Uranus and Neptune. Neptune has been under observation for only a small part of a revolution, so that data thus far obtained seem to ma..y astronomers insufficient for the purpose. Yet a number of astronomers have sought by calculation to prove the existence of such a planet. While their results are discordant, yet they indicate very closely the regions of the sky where search for the hypothetical body might be rewarded with success.

Professor W. H. Pickering, of Harvard, in 1909 evolved a method for the discovery of the distant planet, a method which he first tested by application of the data available to Adams and Leverrier for the discovery of Neptune, and found that the method would succeed. Proceeding then with his hypothetical planet, which he termed "O," he found that it was 51.9 times as far distant from the Sun as the Earth, tho its mass was but twice that of this planet, and that it had a period of revolution of 373.5 years. The problem presented by Uranus, Neptune and "O," according to Professor Pickering, is quite the same as that of Mercury, Venus and the Earth, which has been thoroly studied, so that the relative motions are well understood.

But in investigating the effect that such a hypothetical planet would have on the motion of Uranus, Professor A. Gaillot recently arrived at the conclusion that there were indications pointing to the possibility of still another and more distant planet also exercising a perturbing influence. The results of his calculations and studies therefore indicate the possible existence of two ultra-Neptunian planets, one at a distance from the Sun equal to 44 times the Earth's mean distance and having a mass about $\frac{1}{9000}$ the mass of the Sun, the other having a distance 66 times that of the Earth and a mass about $\frac{1}{11000}$ that of the Sun. While these figures disagree with those of Professor Pickering, yet the position calculated for the second planet agrees quite closely with that of the Harvard astronomer. The problem is by no means solved. It is mentioned to show that a plausible case has been made out for at least one ultra-Neptunian planet.

After astronomers had definitely decided how far the planets and the Sun are situated from the Earth and how they move with respect to one another, they began to wonder if perhaps the whole solar system did not in turn revolve around some central orb. The possibility first occurred to Tobias Mayer (1726), John Michel (1767) and Joseph Jérome Le François Lalande (1776), but the problem was not attacked with any degree of success until Sir William Herschel in 1782 began to draw conclusions from his study of the Milky Way and decided that the entire solar system was drifting toward the constellation Hercules. But Herschel's theory did not meet with general acceptance for many years. Other astronomers suggested various stars as possible central suns controlling the movement of our Solar System. Thus Mädler not only proposed that Alcyone, the principal star in the Pleiades, should be the central sun, because of its situation at a point of neutralization of opposing tendencies and consequently at rest, but even went so far beyond the limits of astronomy as to declare that "Here was the seat of the Almighty, the Mansion of the Eternal." It is hard even for science to quell the imagination and to confine an observer to facts. Mädler's theory was short-lived. Further study of the stars demonstrated the soundness of Herschel's views.

When a modern telescope is turned toward the Milky Way this white girdle of the celestial sphere is resolved into a vast number of stars of which more than 140,000,000 already have been counted on photographs. Each of these stars is a sun like that which governs the Earth, probably surrounded by planets like the Earth, and all these solar systems also are moving, many of them more swiftly than ours.

It was inferred from Herschel's measurements of stellar positions, distances and motions that the solar system was situated comparatively near the center of a universe shaped like a thin, double-convex lens. This universe was supposed to rotate as a unit about its center, with the result that the Sun (comparatively near that center, but absolutely at an immense distance from it) moved in a circle of dimensions so vast that since the discovery of its motion it had not deviated appreciably from a straight line, but had steadily directed its course toward the constellation Hercules.

This simple scheme must now be abandoned, for it has

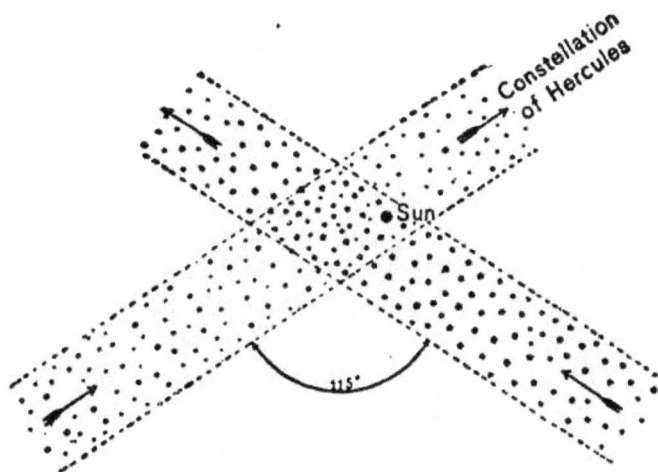

Fig. 16 —The Place and Destination in the Universe of the Solar System.

been discovered by Professor J. C. Kapteyn, a Dutch astronomer, that the visible universe consists of two distant parts. The scientific imagination is compelled to picture two processions of stars moving in paths which make an angle of 115 degrees with each other. One of these stellar streams moves three times as fast as the other. The Sun forms a part of one of the streams and is at present at their intersection.

Altho it is known in a general way that the entire solar system is moving through space at the rate of 12 miles a second, the shape and size of the Sun's orbit are utterly unknown. The changes of environment, accordingly, that will accompany the description of it defy conjecture. The actual course of the solar system is inclined at a small angle to the plane of the Milky Way. Presumably it will become deflected, but perhaps not sufficiently to keep the system clear of entanglement with the galactic star-throngs. In the present ignorance of their composition, no forecast of the results can be attempted; they are uncertain and ex-orbitantly remote. Hence, in a sense, the world knows where it is and in what direction it is moving. But what is the goal, when shall it be reached and what will happen then? Or, in this crossing of the congested thoroughfares of the heavens, will the world be shattered by a collision and resolved into a glowing nebula? This has been the fate of many stars, of several within the period of human history and of one—Nova Persei—within a few years and witnessed in all its destructive detail by astronomers still living.

CHAPTER IX

THE preëminence of the Sun among the celestial bodies and its obvious importance to the life of the world has given to it a unique place, not only in astronomy but in philosophy and religion. The rising of the Sun and the light that it casts over the Earth are distinctly symbolical of the conquest of light over darkness or the triumph of truth over error, so that in many schemes of mythology the Sun god assumes the highest rank and rules over the other elements, such as the Moon, the rains and floods and the stars. Of this there are abundant instances in all primitive religions. Among the Oriental nations, to whom we owe the idea of a flood from which the world emerged, the Sun god reigned supreme. In Egypt, among the Chaldeans and Babylonians and later among the Greeks and Romans the Sun held like sway. In Persia, Sun worship was developed into a more formal religion, which survived for many years.

After the mind of man had developed to a point where it was concerned with philosophical speculation rather than with the mysticism handed down by the priesthood, there came a desire to understand the relation of the Sun to the Earth and to the universe. Its motion across the sky was obvious as well as its change of position from time to time. On this, as has been seen, various theories were founded. It was early realized that the Sun described an annual path on the celestial sphere, which path is a great circle, and this great circle, known as the ecliptic because eclipses take place only when the Moon is in or near it, is at an angle to the equator of the sphere. This angle is termed the obliquity of the ecliptic and was measured by the Chinese, it is claimed, as early as 1100 B.C. with remarkable accuracy. Later the same feat was claimed for Pythagoras or Anaximander in the sixth century B.C., both of whom probably derived their information from the Chaldeans or Egyptians.

When the Sun crosses the equator the day is equal to the night, which occurs twice a year, at the vernal equinox about March 21 and the autumnal equinox about September 23. When the Sun is at its greatest distance from the equator on the north side the time is known as the summer solstice, and when at its greatest distance on the south side it is termed the winter solstice. The positions of these points in the heavens were also known to the early Chinese astronomers with considerable accuracy, while Anaximander, who supposed the Sun to be of equal magnitude with the Earth, used a gnomon or vertical pillar casting a shadow to observe the solstices and equinoxes.

Anaximenes is said to have believed that the Sun was a mass of red-hot iron or of heated stone somewhat larger than the Peloponnesus. He looked upon the heavens as a vault of stones, prevented from falling by the rapidity of its circular motion, while the Sun could not proceed beyond the tropics on account of a thick and dense atmosphere which compelled it to retrace its course. Later Philolaus of Crotona, who was a disciple of Pythagoras and followed his master's teaching that the Earth revolved about the Sun, assumed that the Sun was a disk of glass which reflected the light of the universe. Eudoxus of Cnidus, about 370 B.C., stated that the diameter of the Sun was nine times greater than that of the Moon, which marked a triumph over the illusions of sense. About the time of Alexander, the Great Pytheas, using the gnomon, determined the length of the shadows cast at the summer solstice in various countries. His observations are the most ancient of those preserved after those made in China.

The study of the Sun was undertaken very systematically by Hipparchus, who discovered that the solar orbit was eccentric and that the Sun moved at different speeds at different parts of its journey. With his observational data Hipparchus produced the first tables of the Sun which are mentioned in astronomy. He was enabled to determine the difference between the solar day or time as shown by the Sun and the time indicated by some such measuring device as the clepsydra or water-clock.

The motion of the Sun was also studied by Ptolemy. He compiled solar tables more extensive than those of Hipparchus, which were employed until Albategni (b. 815) made a new compilation of greater accuracy, and which served as a connecting link between the astronomers of Alexandria and those of modern Europe. The obliquity of the ecliptic was constantly being studied. Ulugh Begh, a Tartar prince and grandson of the great Tamerlane, at Samarkand, using a gnomon 100 feet in height, determined the obliquity of the ecliptic or precession of the equinoxes, and secured data for the construction of astronomical tables which were of considerable accuracy.

The apparent motions of the Sun furnished many difficult problems for the astronomer, since the observational data lacked accuracy on account of the absence of satisfactory instruments. Measuring angles by the shadow cast and positions in the sky by crude forms of angular measurement were not adequate for exact work. Not until the advent of the quadrant and the telescope with its various adjuncts was scientific measurement possible. But there were from time to time solar eclipses, of which a careful record was maintained and which the ancient priests noted in connection with their calendar observations. These eclipses played a most important part in ancient astronomy.

An eclipse of the Sun occurs when Earth, Moon and Sun are in direct line at the time of new Moon. As the latter lies between the Earth and the Sun its dark body will pass across the Sun's disk, cutting off the direct illumination. If the Earth cuts off the sunlight from the Moon there is a lunar eclipse. Solar eclipses are of three kinds, partial, annular and total. In the first the Moon, instead of passing directly between the Earth and the Sun, slips past on one side and cuts off from sight only a portion of the Sun's surface. In the annular eclipse the Moon is centrally interposed between the Sun and the Earth, but falls short of the apparent size required to conceal the solar disk entirely. Consequently at the height of obscuration a bright ring is visible around the Moon. In a total eclipse, however, the Sun completely disappears behind the dark body of the Moon. The difference between a total and an annular eclipse depends upon the fact that the apparent diameters of the Sun and the Moon are so nearly equal as to preponderate alternately one over the other through the slight periodical changes in their respective distances from the Earth.

It is the total eclipse that particularly arouses the attention of astronomers, for it cuts off entirely the light of the Sun, and in addition to enabling the observation of the various parts of its surface as it passes across its disk to be made, also makes it possible for an observer to see the stars and planets in the daytime even if they are very near the Sun. A total eclipse of the Sun is not visible on the entire Earth, but only along a comparatively narrow band, lying roughly from west to east and measuring about 165 miles in width. A partial eclipse is seen for about 2,000 miles on either side of this band, but otherwise the phenomenon is not visible on the Earth's surface.

Chinese records going back over 4,000 years describe a solar eclipse which occurred during the twenty-second century B.C. That eclipse carried with it a distinct moral lesson. Two bibulous state astronomers, Ho and Hi, unfortunately happened to be drunk on the day of its occurrence and hence incapable of supervising the performance of the required rites, which consisted in beating drums, shooting arrows and other ceremonies intended to frighten away the mighty dragon who was about to swallow up the Lord of Day. Altho the eclipse was only partial, nevertheless great confusion resulted, and Ho and Hi were put to death as a lesson to later astronomers.

Another early record of a total eclipse comes from Babylon, 1063 B.C. Several centuries later Assyrian tablets record solar eclipses. Herodotus, Plutarch and the Bible all refer to the phenomenon. Thus one is enabled to determine with accuracy the time of historical events. Likewise in the Anglo-Saxon chronicles several notable eclipses are recorded.

The Sun appears as a brilliant white body. Just as the Earth is enveloped by an atmosphere, so the Sun is surrounded by several layers of gaseous and vaporous matter, with the result that so far as an observer on the Earth is concerned its nucleus is quite invisible. These layers are more or less transparent, just as the atmosphere of the Earth is transparent, so that the bright white body of the Sun is visible only through these various envelopes. This bright white portion is called the photosphere and is the source of the light and heat which is radiated to the Earth. Here are found the sun-spots distinctly seen with the telescope and even by the naked eye. Under the photosphere it may be that the more solid portions of the Sun are situated, but it is obvious that its surface consists of highly incandescent vapors above which is a smoke-like haze. Upon this rests the reversing layer, which is composed of glowing gases, but is cooler than the photosphere beneath. It has a thickness of between 500 and 1,000 miles and contains, as the spectroscope shows, many of the elements of which the Earth is composed, in the form of vapor. Above the reversing layer comes the chromosphere. The chromosphere is between 5,000 and 10,000 miles in thickness and is composed of glowing gases, chief among which is hydrogen. The chromosphere is a brilliant scarlet in hue, but the redness is entirely overpowered by the intense white light of the photosphere, which shows through from behind. The most interesting features of the chromosphere are the red prominences, which are the marks of violent agitation in its upper portions and which

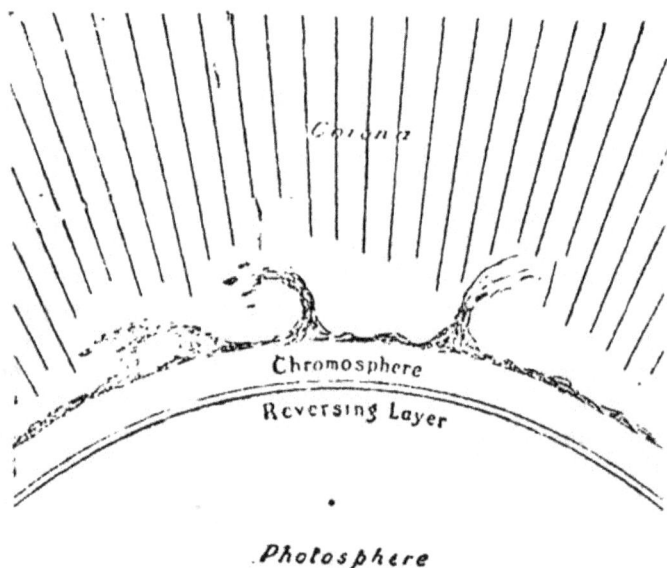

Photosphere

Fig. 17 —THE SOLAR STRUCTURE.

are such a notable feature in total solar eclipses. After the chromosphere comes the corona, which is the outer envelope of the Sun and consists of a halo of pearly white light of irregular outline which fades away into the surrounding sky and extends outward for many millions of miles. The corona suffers so much from the brilliancy of the photosphere that it is only on the occasion of a total solar eclipse that it can be seen in all its remarkable beauty.

As the photosphere, the reversing layer and the chromosphere are all sources of light, the solar spectrum observed in spectroscopes is composed of the three separate spectra combined. For this reason eclipses afford welcome opportunities for studying the Sun's surface. When the Moon completely covers the photosphere its brilliant light is cut off and the other features of the Sun can be examined visually, and, what is more important, spectroscopically and photographically. Thus when the spectroscope is directed to the reversing layer during a total eclipse the dark lines of the solar spectrum change into bright lines or are reversed. But this reverse spectrum is a phenomenon of a moment only, and as the Moon progresses an altered spectrum is obtained. That of the chromosphere is of sufficiently long duration to permit an estimate of its depth and nature, and finally, when this is covered up, there is the corona, which has a distinct spectrum of its own, containing a strange line, the distinguishing green of which has not yet been identified with any element known upon the Earth.

Modern conceptions of the Sun are due very largely to the use of the telescope, the spectroscope and the spectroheliograph, especially the last two instruments. With the telescope, when the intense light of the Sun is properly reduced, it is possible to examine its surface and obtain a certain amount of information as to its nature or to obtain photographs of that surface by very short exposures. But on the spectroscope and the spectroheliograph the astronomer depends for his knowledge of the constitution and composition of the great center of the solar system. The development and nature of the spectroscope, as used in solar research, have been already discussed, but it is appropriate to add here a brief explanation of the spectroheliograph, for to its use is due not only a large part of present-day information of the prominences, but more recently an explanation of the sun-spots themselves and the study of various features of the photosphere.

The spectroheliograph was first devised in successful working form by Prof. George E. Hale at Kenwood Observatory, Chicago, in 1889. "The principle of this instrument is very simple," writes Professor Hale. "Its object is to build up on a photographic plate a picture of the solar flames by recording side by side images of the bright spectral lines which characterize the luminous gases. In the first place, an image of the Sun is formed by a telescope on the slit of a spectroscope. The light of the Sun after transmission through the spectroscope is spread out into a long band of color, crossed by lines representing the various elements. At points where the slit of the spectroscope happens to intersect a gaseous prominence the bright lines of hydrogen and helium may be seen extending from the base of the prominence to its outer boundary. If a series of such lines corresponding to different positions of the slit on the image of the prominence were registered side by side on a photographic plate, it is obvious that they would give a representation of the form of the prominence itself.

"To accomplish this result it is necessary to cause the solar image to move at a uniform rate across the first slit of the spectroscope, and, with the aid of a second slit (which occupies the place of the ordinary eye-piece of the spectroscope), to isolate one of the lines, permitting the light from this line and from no other portion of the spectrum to pass through the second slit to a photographic plate. If the plate be moved at the same speed with which the solar image passes across the first slit, an image of the prominence will be recorded upon it." The same method answers for the study of the sun-spots and other features of the Sun's surface. As the result of telescopic, spectroscopic and spectroheliographic observation it is now known that the Sun's principal features are its sun-spots, its photosphere, its chromosphere and its corona.

A sun-spot when examined through a telescope consists of a dark central region called the umbra into which long.

narrow filaments reach. The space occupied by these filaments is termed the penumbra. The darkness of the umbra is not absolute, but is relative to the more brilliant surface of the photosphere, and if observed alone would be far more brilliant than the most powerful arc light. These dark spots on the Sun were familiar objects in the days of pretelescopic observation. But their importance to astronomy dates with their discovery in 1610 by Galileo with his telescope. The great Italian astronomer did not announce his discovery until May, 1612, by which time sunspots had been seen independently by Thomas Harriott (1560-1621), John Fabricius (1587-1615), who published his observations in June, 1611, or before Galileo, and the Jesuit Father Christopher Scheiner (1575-1650), all of them pioneer observers with the telescope.

Before sun-spots were clearly observed with the telescope it was assumed that they were due to the transit of Mercury. Even Father Scheiner, after his telescopic studies, suggested that the spots might be small planets revolving around the Sun and appearing as dark objects whenever they passed between the Sun and the observer. It was recognised, however, that the spots appeared to move across the face of the Sun from the eastern to the western side, or roughly from left to right. Father Scheiner's view was also held by Jean Tarde, Canon of Sarlat; by Father Malpertius, a Belgian Jesuit, and later by William Gascoigne, the inventor of the micrometer. Galileo, however, advanced a cloud theory, while Simon Marius, "astronomer and physician" to the brother Margraves of Brandenburg, proposed the ingenious "slag theory," according to which the dark spots were the cindery refuse of a great solar conflagration and occasionally expelled in the form of comets, which afterward blazed up with renewed vigor. Galileo in a controversy with Father Scheiner proved him wrong in his planetary theory, while the occurrence of comets in 1618 won supporters for the theory of Marius. Galileo also ascertained the rotation of the Sun in a period of between 25 and 26 days, as well as the general zone of the sun-spots.

The next important contribution to sun-spot theory came from Derham, whose observations were made during the years 1703-1711. He believed that the spots on the Sun were caused by the action of some new volcano, whose smoke and other "opacous matter" produced the spots. As they decayed they became half shadows encircling the darker portions and finally became bright spots. Lalande, the celebrated French astronomer, believed that the spots were rocky elevations, about which the penumbra represented shoals or sandbanks, while around flowed enormous oceans. Lalande's explanation, as well as that of Derham, was clearly based upon terrestrial analogies, which were employed with considerable frequency by early astronomers.

In 1769 Alexander Wilson, of Glasgow (1714-1786), examining the large sun-spot visible that year, noted that, as the Sun's rotation carries a spot across its disk, there was a change in its appearance, and that the same effect of perspective was produced as if it were a saucer-shaped depression, the bottom forming the umbra or central black spot and the sloping sides the penumbra or surrounding portion of half shadow. The penumbra appeared narrowest on the side nearest the center of the Sun and widest on the part nearest the edge. Hence Wilson assumed "that the great and stupendous body of the Sun is made up of two kinds of matter, very different in their qualities, that by far their greater part is solid and dark, and that this immense and dark globe is encompassed by a thin covering of that resplendent substance from which the Sun would seem to derive the whole of his vivifying heat and energy." Wilson went on to explain that the excavation of spots might be occasioned "by the working of some sort of elastic vapor which is generated within the dark globe," and that the luminous material, which was more or less fluid, was acted upon by and tended to throw down and cover the nucleus.

Sir William Herschel devoted considerable attention to the Sun, and observing the variation in the sun-spots, reached the conclusion in 1801 that it indicated a certain variability in the total amount of solar radiation, which he assumed might have some connection with terrestrial phenomena, especially the weather. He endeavored first in 1801 to trace connection between the price of wheat, naturally influenced by the effect of weather on the crops, and the occurrence of sun-spots, claiming that when the latter were scarce there was a diminished solar activity, which caused colder weather, with obvious results. Ingenious as this theory was, there were not sufficient substantiating meteorological data. It finds, however, a counterpart in modern studies of the Sun's heat, not necessarily connected exclusively with sun-spots, however, whereby it is hoped to establish some useful knowledge of the relation between the amount of heat radiated from the Sun and weather conditions on our Earth.

Herschel's observations of the Sun and sun-spots were continued by his son, Sir John Herschel, at the Cape of Good Hope (1836-1837). John Herschel assumed that their motion was due to fluid circulations similar to those producing the trade and anti-trade winds on the Earth. "The spots, in this view of the subject," he said, "would come to be assimilated to those regions on the Earth's surface where for the moment hurricanes and tornadoes prevail, the upper stratum being temporarily carried downward, displacing by its impetus the two strata of luminous matter beneath, the upper of course to a greater extent than the lower, and thus wholly or partially denuding the opaque surface of the Sun below."

Such observation of sun-spots, made with considerable thoreness by the astronomers mentioned, as well as numerous others, did not establish any regularity in their appearance or effacement. It remained for Heinrich

Schwabe (1790-1875) at Dessau to announce in 1843 that the sun-spot phenomena reached a maximum probably in a decennial period. This announcement, altho coming as it did after a 'patient study of the Sun, attracted no particular attention until a series of sun-spot statistics were published in Humboldt's "Kosmos." Then the correctness of Schwabe's observations and deductions was apparent to all. When compared by Dr. John Lamont and Sir Edward Sabine with various periodical magnetic disturbances, it was found that the two cycles of changes agreed with extraordinary exactness. It was a remarkable coincidence that the observations of a number of investigators were in complete harmony. A study of sun-spot records established the decennial period more correctly at 11.11 years. Thus commenced a recognition that magnetic disturbances on the Earth were related in some way to sun-spot phenomena. For many years no direct connection could be established, altho various theories were forthcoming. Likewise further attempts were made to identify the variations of sun-spots with meteorological phenomena, but without success, until Wolf in 1859 by an examination of the Zürich chronicles (1000-1800 A.D.) found data which enabled occurrences of the Aurora Borealis to be correlated with a disturbed condition of the Sun.

From this time on the influence of the Sun on terrestrial conditions assumed new importance. The beautiful phenomenon of the aurora, which consists of a glow in the sky about the north and south poles, had been observed for ages, but the first scientific connection of importance recorded was in 1716, when Halley stated that the Northern Lights were due to magnetic "effluvia." In 1741 Hiorter at Upsala observed that they produced an agitation of the magnetic needle. This connection was further demonstrated by Arago (1819), so that by the middle of the nineteenth century the connection of the Aurora with the Sun and in turn with terrestrial magnetism was as evident as it was insufficiently explained.

The first result of the modern study of the sun-spots was to put an end to the old notion that there was a dark and cold interior of the Sun and that the sun-spots were merely rents in the brilliant cloud covering through which the interior portion could be seen. The late Prof. S. P. Langley, one of the most active of the modern students of the Sun and its surface, thought that the filaments which, taken together, constitute the penumbra, were everywhere present on the surface. Professor Hale states: "He regarded them as resembling the stalks of a wheat-field, seen on end in the undisturbed photosphere, and revealing more of their true characteristics in the penumbra, where they are bent over and drawn out toward the central part of the spot. Langley believed that we are observing clouds of luminous vapors rising from the Sun's interior, the seats of convection currents which bring to the surface the immense supplies of heat radiated by the Sun into space. Separating these luminous columns are darker regions, characterized by a lower degree of radiation.

"The minute details can be recorded only with the greatest difficulty. Under ordinary atmospheric conditions the solar image is not seen as a sharp and well-defined object, but its details are continually blurred by the effect of irregularly heated currents in our atmosphere. Even under the best conditions the moments of very sharp definition are few, and the greatest patience and perseverance are required on the part of an observer who would record his impressions of the solar structure. At the best, drawings based upon visual observations must be unsatisfactory, since even the skilled hands of Langley could not secure the perfect precision which is so desirable. It accordingly might be hoped that here, as in other departments of solar research, photography would afford the necessary means of securing results unattainable by the eye. Unfortunately, however, this hope has been only partially realized."

The influence of sun-spots is not confined to magnetic and electrical phenomena. The researches of Köppen,

which have been confirmed by Newcomb, show that the
average temperature of the Earth determined by the combination of a great number of thermometric observations
made at several stations indicated a fluctuation of 0.3° to
0.7° C. during the 11-year sun-spot period. In other words,
the temperature of the Earth's atmosphere indicates small
fluctuations which correspond with the sun-spot period,
thus indicating that the solar heat radiation varies with the
number of the sun-spots. The mean temperature of the
Earth is greatest at the time of minimum sun-spots and
lowest at time of maximum sun-spots. Hence the determination of the amount of heat radiated by the Sun at various times, especially at sun-spot maxima and minima, is
a matter of considerable terrestrial importance.

The study of the sun-spots carried on by Professor Hale
with the spectroheliograph and other apparatus, including
special red-sensitive plates of considerable speed, reached
an interesting stage in 1908, when it was demonstrated
that sun-spots are centers of attraction which draw toward
them the hydrogen of the solar atmosphere. Subsequently
it was found that these spots are the seats of great cyclones,
in which cool hydrogen gas is set whirling and is sucked
down in the great maelstrom of the Sun, rushing into the
center of the spot at a rate of about 60 miles a second.
Consequently the spots are the center of great solar disturbances which are of an electro-magnetic nature. According to the modern electronic theory of matter and
electricity, "electrons," or minute particles of matter, in
their terrific cyclonic velocity produce magnetic lines of
force. It was found by Professor Zeeman that, when light
is passed through a strong magnetic field, the lines of the
spectrum are subdivided and appear double. This Zeeman
effect Professor Hale has found in the spectrum of the
sun-spots. If one looks at the center of a spot the light
travels in the direction of the axis of the whirl or cyclone,
while in viewing a spot at the edge of the Sun the direction is at right angles to the axis and is manifested accord-

ingly in the spectrum. This theory has been thoroly confirmed, so that to-day it is known that sun-spots are magnetic fields of great intensity. The important discoveries made by Professor Hale and his associates may thus be summarized as follows: First, that the spots are cooler than the surrounding region;.second, that they are centers of violent cyclones, and, third, that they are magnetic fields of great intensity.

In addition to the sun-spots the photosphere includes other interesting features, notable among which are the 'faculæ,' "little torches," so named by Father Scheiner. These bright, globular objects besides the sun-spots, are the only other phenomena of the Sun's surface visible by direct observation. Schröter showed that the faculæ are heaped-up ridges of the disturbed photospheric matter. Secchi and Young assumed that the faculæ are the result of violent eruptive action of the sun-spots, but it remained for the spectroheliograph to give a clear idea of their nature. Professor Hale states that "they are usually most numerous in the vicinity of sun-spots, and near the Sun's limb they are sometimes very conspicuous brilliant objects covering large areas. Near the center of the Sun, however, they are practically invisible, tho faint traces of them can sometimes be made out on photographs taken with a suitable exposure. This increase of brightness toward the Sun's limb is assumed to be due to the elevation of the faculæ above the photospheric level and their escape from a considerable part of that absorption which so materially reduces the brightness of the photosphere. Rising above the denser part of the absorbing veil and thus suffering but little diminution of light, they appear near the Sun's limb as bright objects on a less luminous background. The chief difference of the faculæ from the rest of the photosphere lies in their greater altitude, as photographs have shown that they may be resolved into granular elements similar to those constituting the photosphere. But they are the regions from which immense masses of vapors rise to the solar surface and for that reason are important in the solar mechanism.

"Near the edge of the Sun their summits lie above the lower and denser part of that absorbing atmosphere which so greatly reduces the Sun's light near the limb, and in this region the faculæ may be seen visually. At times they may be traced to considerable distances from the limb, but as a rule they are inconspicuous or wholly invisible toward the central part of the solar disk. The Kenwood experiments had shown that the calcium vapor coincides closely in form and position with the faculæ, and hence the calcium clouds were long spoken of under this name. In the new work at the Yerkes Observatory the differences between the calcium clouds and the underlying faculæ became so marked that a distinctive name for the vaporous clouds appeared necessary. They were therefore designated flocculi, a name chosen without reference to their particular nature, but suggested by the flocculent appearance of the photographs."

"With the spectroheliograph," Professor Hale relates, "it was at once found possible to record the forms, not only of the brilliant clouds of calcium vapor associated with the faculæ and occurring in the vicinity of the sun-spots, but also of a reticulated structure extending over the entire surface of the Sun. . . . From a systematic study of spectroheliograph negatives, in the course of which the heliographic latitude and longitude of the calcium clouds, or flocculi, in many parts of the Sun's disk were measured from day to day (by Fox), a new determination of the rate of the solar rotation in various latitudes has been made. This shows that the calcium flocculi, like the sun-spots, complete a rotation in much shorter time at the solar equator than at points nearer the poles. In other words, the Sun does not rotate as a solid body would do, but rather like a ball of vapor, subject to laws which are not yet understood."

CHAPTER X

THE so-called "reversing layer" was discovered by the
late Professor Charles A. Young during the eclipse of
December 22, 1870, on which occasion he placed the spectroscope with its slit tangential to the Sun's limb, so that
it ran along a shallow bed of incandescent vapors. When
the Moon reduced the size of the crescent of the Sun the
dark lines of the spectrum and the spectrum itself gradually faded away until all at once, as suddenly as a bursting
rocket shoots out its stars, the whole field of view was
covered by bright lines more numerous than one could
count. This phenomenon lasted for about two seconds and
gave the impression of a distinct reversal of the Fraunhofer spectrum, showing bright lines for dark in every
case. That such a reversing layer should exist was demanded by Kirchoff's theory of the production of the
Fraunhofer lines, and implied a stratum of mixed vapors at
a lower temperature than that of the surface of the Sun.
It was by such a stratum that the missing rays of the solar
spectrum were stopped. The spectrum from this portion
alone should supply bright lines if the overpowering brilliancy of the solar background could be cut off, which can
occur only at the time of an eclipse. This observation of
Professor Young's, with its important bearing on the theory of Kirchoff, was not confirmed until 1896, when photo-

graphic evidence was forthcoming. During the eclipses of 1898 and 1900 abundant corroborative material was obtained, and the reversing layer as a reality was conclusively demonstrated. The total depth of this reversing layer has been placed at from 500 to 600 miles. It continues in a normal state of tranquillity, for little change is produced in the aspect of the dark lines.

The chromosphere or envelope of glowing gases which covers the Sun completely was detected by observers of eclipses in the eighteenth and nineteenth centuries. There is a record in a letter from Captain Stannyan to Flamsteed, the British Astronomer Royal, describing an eclipse witnessed at Berne on May 1 (O. S.), 1706, in which he states that the Sun's "getting out of the eclipse was preceded by a blood-red streak of light from its left limb." Halley and De Louville in 1715 noted a similar phenomena, and it was also observed during annular eclipses of 1737 and 1748, but with the ruby brilliancy toned down to "brown" or "dusky red" by the surviving sunlight. During the eclipses of 1820, 1836 and 1838 similar observations were made, but it was not until the eclipse of the 8th of July, 1842, that the virtual discovery of the chromosphere as a solar appendage may be said to have been made. The eclipse of 1868, which was observed spectroscopically and photographically as well as with the telescope, served to make clear the nature of the chromosphere and to reveal that it is a continuous envelope of hydrogen and other incandescent gases, some thousands of miles in thickness and of the same eruptive nature as the prominences which are shot out from it. In other words, it seems to be a collection of minute flames set close together and giving it the appearance of a large conflagration. The summits of the flames of fire, which incline when the Sun's activity is greatest, are erect during its phase of tranquillity.

The chromosphere is marked by an irregular distribution over the Sun's surface, which in no way partakes of the character of an atmosphere. Professor Hale in 1897 discovered a low stratum of carbon vapor. Such rare metals as gallium and scandium have been discovered with the spectroscope. The vapors of magnesium, iron and several other substances are conspicuously represented in the spectrum of the chromosphere, and with the Yerkes telescope the fine bright lines due to the vapor of carbon also may be seen.

The solar prominences are conspicuous eruptive or flame-like emanations from the chromosphere which are seen at total eclipses of the Sun. They project like red flames beyond the dark edge of the Moon, and were first described by Lector Vassenius, a Swedish professor of Gothenburg, who observed the total eclipse of May 2 (O.S.), 1733. One of these reddish clouds outside of the solar disk was so large that it could be detected with the naked eye. The phenomenon excited his admiration and wonder. The prominences were also observed in 1778 by the Spanish Admiral Don Antonio Ulloa, who was convinced of their connection with the Sun on account of their color and magnitude. By some observers the solar prominences were regarded as the illuminated summits of lunar mountains, and by Arago they were described as solar clouds shining by reflected light. Abbé Peytal, in 1842, spoke of them as self-luminous and as a third or outer solar envelope composed of the glowing substance of the bright rose tint which produced mountains, just as clouds were piled above the Earth's surface. In 1851 Hind, an English astronomer, noted on the south limb of the Moon "a long range of rose-colored flames," which Dawes spoke of as a low ridge of red prominences resembling in outline the tops of a very irregular range of hills. Airy also noted this rugged line of projections, and spoke of its brilliancy and "nearly scarlet" color. But the truly solar origin of the phenomena was not conclusively demonstrated until 1860, when the prominences were photographed by Secchi and De la Rue, and shown to be independent of the motion of the Moon. In 1868, with the growth of spectroscopy and solar chemistry, the gaseous nature of the prominences and their connection with the Sun was made evident. They were found to consist of immense masses of hydrogen and helium gas rising from the chromosphere and reaching an altitude of hundreds of thousands of miles.

The Corona.—The corona is a beautiful lustrous solar wrapping, which can be observed only during a total eclipse. The winged circle, the winged disk, or the ring with wings, as it is variously called, found upon Assyrian and Egyptian monuments, may be reproductions of the phenomenon. The first definite mention of a solar corona is to be found in Plutarch, in connection with the eclipse which probably took place in 71 A.D. He writes that the obscuration caused by the Moon "has no time to last and no extensiveness, but some light shows itself around the Sun's circumference which does not allow the darkness to become deep and complete." Kepler mentions a ray of light seen around the eclipsed Sun in 1567, and ascribes it to some sort of luminous atmosphere around the Sun. In 1706 Cassini, observing an eclipse of the Sun in France, saw the "crown of pale light" around the lunar disk, and stated that it was caused by the illumination of the zodiacal light. Halley, observing an eclipse in London in 1715, describes minutely the phenomena of the luminous ring rising around the Moon to a great height, and showing considerable brilliancy. The eclipse of 1842 was the first to indicate the corona's great importance to astronomers, and from that date it received careful attention and earnest study.

In 1869 Professor Harkness and Professor Young discovered a bright line of unknown origin in the coronal spectrum, showing that it consists in large part of glowing gases. With the advent of astronomical photography, and with the development of the spectroscope, more attention than ever was given to the careful study of the corona in the limited time available on the occasion of a total eclipse. The corona is described by Professor Hale as a "faintly luminous veil of light extending outward in long streamers from the surface of the photosphere to distances of several millions of miles, and exceeded in brilliancy, even in its brightest parts, by the full Moon. In many ways its streamers resemble those of the Aurora Borealis, and it is indeed possible that their origin may be ascribed to some similar electrical cause. During the few minutes of a total eclipse they are not seen to undergo change of form, but the outline of the corona does vary greatly from year to year, in sympathy with the general variation of the solar activity.

"Spectroscopic observations have shown that the corona consists mainly of gases unknown to the chemist. That is to say, the lines in its spectrum do not coincide in position with the lines of any terrestrial element. Whether these gases, which are probably very light, will ultimately be found on the Earth cannot be predicted. Like helium, first known in the Sun, they may eventually be encountered in minute quantities in some mineral where they have hitherto escaped the chemist's analysis. The fact that the lower part of the corona gives a continuous spectrum, with a feebler solar spectrum superposed upon it, indicates that minute incandescent particles are present, which are hot enough to radiate white light, and which scatter enough sunlight to account for the presence of the solar spectrum." The strange line in the green portion of the spectrum does not correspond with that of any element with which we are acquainted upon the Earth, and accordingly a hypothetical element has been assumed, to which the name of "coronium" has been applied.

Chemical Composition.—Anaximenes' idea that the Sun was a glowing ball of incandescent iron came almost as near the truth as subsequent speculations by philosophers and astronomers, until about the middle of the nineteenth century. In 1859 Kirchoff's great discovery of the explanation of the Fraunhofer lines in the solar spectrum made it possible to ascertain the chemical compositon of the Sun. Thus, as has been shown, the bright-colored band formed of light from a small hole in Newton's shutter passing through a prism made that prism the forerunner of an instrument able to teach the nature and constitution of bodies far distant in the heavens. Thomas Melvill had examined with a prism various flames in which different substances had been introduced, and had reached the conclusion by the middle of the eighteenth century that certain vapors, notably sodium, contained light which had a definite place in the spectrum. Melvill's deep yellow ray became the sodium line in the spectroscope of Fraunhofer.

When Kirchoff noted the identity of certain lines characteristic of terrestrial elements, and then assumed their presence in the Sun, he laid the foundation of the modern science of solar chemistry. Wherever light can be obtained from a heavenly body it is now possible to resolve it into its spectral elements and thus to identify the substances. In fact, as substances such as helium were found in the Sun by Lockyer, which had no terrestrial counterpart, hypothetical elements were assumed. The spectroscope has shown in the Sun the presence not only of gases such as hydrogen and helium, but iron, sodium, magnesium, calcium, and many other substances. Hence the chemical composition of the Earth and the Sun are much the same, altho there is evidence of the existence in the Sun of substances not yet found in the Earth. When the spectroscope was applied to the analysis of the chromosphere and its prominences it was found that they are composed of the vapor of calcium and of the light gases helium and hydrogen. Sun-spots, too, have been found to have a characteristic chemical composition, while the corona emits rays which probably indicate the presence iu it of very light and tenuous gases.

CORONA OF THE SUN, TAKEN DURING TOTAL ECLIPSE. (Yerkes Observatory.)

With the light given off by the Sun there is to be considered a phenomenon which only recently has been demonstrated by experiment, namely, that light exerts a pressure which can act on minute particles quite as effectively as gravitation. Gravitation, however, attracts entire masses, but pressure acts only on surfaces, so that a force such as the pressure of light, to be effective, must deal with very minute masses. If you subdivide a mass into a large number of minute particles the effect of gravitation is not changed, but a point will be reached in the subdivision where particles may be obtained having much surface and very little weight. If such a particle has a diameter of $1/100000$ of an inch it will be exactly balanced in space, pulled by gravitation (weight) on the one hand, pushed by light on the other. If the particle is even smaller than $1/100000$ of an inch in diameter the pressure of light upon it pushes it away with terrific force. It is the radiation of the Sun acting on the minute particles that produces the phenomenon of comets' tails, and it is this pressure which may be responsible for the brilliant phenomena of the corona, visible only during the vanishing moments of a total eclipse. No one has ever satisfactorily explained how the highly attenuated matter composing both the prominences and the corona is supported without falling back into the Sun under the pull of solar gravitation. Now that Arrhenius has cosmically applied the effects of light pressure a solution is presented.

How difficult it is to account for such delicate streamers as the "prominences" on the Sun is better comprehended when we fully understand how relentlessly powerful is the grip of solar gravitation. The Sun admittedly projects vapors into space, vapors which must condense into drops when they encounter the cold of outer space. If the drops are larger than the critical size which determines whether light-pressure or gravitation shall prevail they will be snatched back by the Sun's gravitational attraction and give rise to the curved prominences that are often observed. If they have approximately the critical diameter they will float above the Sun in the form of beautiful carmine clouds, balanced in space by the equal and oppositely acting forces of gravitation and radiation pressure. These clouds have hitherto been particularly puzzling, for in the absence of a dense solar atmosphere their existence seemed a celestial paradox. If the condensed drops are smaller than the critical diameter they will be projected by the pressure of light far beyond the Sun, to form the beautiful pearly corona.

From the fact that comets have passed through the corona without any very apparent retardation, some idea of its tenuity may be gained. Assuming that the corona consists of particles of such size that the radiation pressure on each exactly equals its weight, Arrhenius finds that the entire corona weighs no more than 12,000,000 long tons, which is equivalent to four hundred large transatlantic steamers, and is not more than the amount of coal burned on the Earth every week.

Compared with the infinity of the space in which it is poised, the Earth is smaller than a vanishingly small speck on a sheet of paper having an area of many square miles. So far as the Earth is concerned, the Sun is very much in the position of a man who throws away all but a single cent of a fortune consisting of twenty-three million dollars; for only $\frac{1}{2300000000}$ of his radiated energy reaches this globe. What, then, becomes of the huge number of corpuscles which are shot from the Sun and which never strike the Earth? It is conceived by Arrhenius and his followers that many of them must collide with corpuscles discharged by suns other than that of this solar system—suns ineffably distant, so that their light reaches the Earth only after the lapse of countless centuries, and so that they are seen not as they gleam now, but as they gleamed when Egypt was young and Greece was a wilderness inhabited by savages. Such collisions must result in the formation of larger masses up to a limit determined by the electrical charges carried by the corpuscles.

Solar Energy —The Earth depends upon the Sun for its supply of heat and light. The Sun is transmitting heat to the Earth, and unless the supply of energy is being replenished in some way it is obvious that it must be losing in heat and temperature. From its effect on the Earth an estimate can be had of the amount of heat radiated by the Sun annually. If it be assumed that the Sun has the same heat capacity as water, and hydrogen is the only substance with a greater heat capacity, it would fall in temperature about 4° F. annually. If, therefore the great luminary were simply a hot body, cooling off, its present rate of radiation could not be maintained for more than 3,000 years. That the Sun has been radiating a much longer period than this is obvious from geological and biological evidence, so that some other cause must be sought.

Up to the nineteenth century the doctrine of infinite durability was generally held by astronomers and geologists. For that reason no particular attention was paid to the nature of the heat of the Sun and its source; but with the formulation of the doctrine of the conservation of energy it was realized that the energy of the Earth must proceed from the Sun for the greater part, and consequently it became necessary in turn to question the source of solar heat. Robert Mayer, to whom is due the earliest conception of the conservation of energy, asked himself: If the Sun is hot, why does it not cool off? In 1848 Mayer published some answers to his own question in a paper which failed to receive the approval of the French Academy of Sciences. His conclusions were as follows: The Sun cannot be a glowing mass sending out radiation without compensation. Solar heat cannot be due entirely to chemical changes, nor can it be due to solar rotation. In his opinion it was the result of meteors

falling into the Sun. He did not overlook the fact that
the resulting increase in the mass of the Sun would increase its attraction for the planets and would shorten
the sidereal year. This was contrary to the facts of observation, and Mayer was forced to an incorrect conception of the undulatory theory of light to explain the situation. Six years later William Thomson, subsequently
Lord Kelvin, reached independently almost the same conclusion as Mayer, but he was able to explain the increase
in the Sun's mass resulting from meteoric showers, for
according to the gravitation theory the "added matter is
drawn from space, where it acts on the planets with very
nearly the same force as when incorporated in the Sun."

Lord Kelvin then ventured an estimate of the age of
the Sun, which was the first attempt in this direction made
by a physicist. He assumed that the solar energy of rotation was derived from the fallen meteors, which, allowing
for the constant loss of solar energy by radiation, could
be acquired in 32,000 years. Taking into consideration
the limited amount of meteoric matter available near the
Sun, he concluded that "sunlight cannot last as at present
for three hundred thousand years." This theory attracted
little attention when promulgated by him in 1854, and
was abandoned by him later.

But in the same year Helmholtz, working along the
lines of the nebular hypothesis of Kant and Laplace, derived the heat of the Sun from the contraction of the
nebula from which the Sun and planets were formed.
He asserted also a further contraction of the Sun, now
assumed to be in progress, by which the kinetic energy
obtained was converted into heat, and compensated for
the loss of solar heat by radiation. Accordingly, if the
Sun contracts one ten-thousandth part of its radius each
year, enough heat would be generated to supply radiation for 2,100 years. Helmholtz' computation gives
twenty-two million years as the probable age of the Sun,
based on a uniform radiation and homogeneous density

of that body. Later, S. P. Langley, with experimental data derived from the direct radiation of the Sun, reduced this age to eighteen million years. This theory immediately supplanted that of the falling meteors, which more serious reflection demonstrated could account only for the slight increase in the solar heat as compared with the energy of shrinkage.

Lord Kelvin, in 1862, returned to the subject, favoring a theory like that of Helmholtz', concluding that "we may accept as the lowest estimate for the Sun's initial heat 10,000,000 times a year's supply at the present rate, but 50,000,000 or 100,000,000 as possible in consequence of the Sun's greater density in his central parts." "As for the future. . . . inhabitants of the Earth cannot continue to enjoy the light and heat essential to their life for many million years longer unless sources now unknown to us are prepared in the great storehouse of creation." Studies of the Sun's heat were continued by Lord Kelvin, and in his theory he incorporated a discovery made in 1870 by J. Homer Lane, an American, which paradoxically demonstrated that within certain limits the more heat a gaseous body loses by radiation the hotter it will become.

This theory of Helmholtz', as modified by Kelvin, encountered a serious rival in 1882 when Sir William Siemens proposed that a rotating Sun hurled by centrifugal action at the equator enormous quantities of gas into space, which returned to it again at the poles, somewhat after the manner of a regenerative furnace. Helmholtz' theory was modified in 1899 by Professor T. J. J. See, who abandoned the German scientist's hypothesis of homogeneous density for the Sun and applied Lane's law, investigating minutely the more complex case of central condensation. As the result of his study the probable age of the Sun was extended from twenty-two million to about thirty-two million years.

All of these researches came before the discovery of

radium, and when the extraordinary properties of this new substance were known it was stated that a bare fraction of a per cent. of radium present in the Sun would account for and make good the heat that is annually lost by radiation. Should this hypothesis hold good, an entirely new aspect is given to the problem of solar heat and to the heat of the Earth. This innovation, advanced by members of the younger school of physicists, all of whom had prosecuted with vigor researches in radioactivity, did not appeal to Lord Kelvin, who maintained in 1906 that the gravitation theory was still sufficient to account for the Sun's heat. No evidence has been produced of the presence of radium in the Sun, altho helium has been found there. As helium is obtained from radium, the existence of radium in the Sun is quite probable.

Sir George H. Darwin, in discussing the effect of these recent discoveries on solar age, says: "Knowing, as we now do, that an atom of matter is capable of containing an enormous store of energy in itself, I think we have no right to assume that the Sun is incapable of liberating atomic energy to a degree at least comparable with that which it would do if made of radium. Accordingly, I see no reason for doubting the possibility of augmenting the estimate of solar heat, as derived from the theory of gravitation, by some such factor as ten or twenty."

It is obvious, therefore, that while the contraction theory explains the origin of a vast amount of the Sun's heat, yet there are other sources of internal energy which recent discoveries plainly indicate are of great importance, so that the scientist at this stage is unable to declare positively the age of the Sun or to make any accurate estimate as to the probable duration of the time through which it will afford light and heat to the Earth and the other planets. Nevertheless, it seems assured that millions of years hence, how many cannot even roughly be determined, the Sun will be reduced from a ball of glowing vapor to a gigantic black cinder rushing through space.

Unwarmed by any central luminary, its crust will be washed by oceans of air liquefied by a cold too intense for any living creature to endure.

The light of the Sun is obviously more intense than any other luminant known to man. If compared with the full Moon, 600,000 times as much light is received from the Sun. Expressed in another way, the Sun gives over 60,000 times as much illumination as a standard candle at a distance of one yard. But not all of the light and heat which is radiated from the Sun comes to the Earth. Professor Langley, in his experiments at Mt. Whitney in 1881, found that a clear atmosphere would cut off 40 per cent. of the rays coming perpendicularly to the Earth's surface. Gases in the atmosphere, such as carbon dioxide, cut off even a greater amount, and the general absorption is greater at the violet end than at the red. It is for this reason that high altitudes and a clear atmosphere are essential for solar investigation; and for this reason, too, that the setting Sun appears red, the bluish rays being absorbed in traversing a greater amount of terrestrial atmosphere than when the Sun is high in the sky.

The temperature of the Sun can be estimated from its brilliancy and from spectroscopic and bolometric studies. It is a common experience that a filament of an incandescent lamp emits more light and glows more brightly when the amount of current is increased. At first the filament is red, but as more current is permitted to flow it becomes yellow, and finally a brilliant white. This is marked by an increase in temperature; and secondly, the temperature depends upon the brilliancy of the glow. The same analogy holds good in the case of the Sun and the stars. If the wave length of the radiation is known, and the color which emits the greatest amount of heat in the spectrum—and this can be measured by the bolometer by a simple law—it is possible to calculate the absolute temperature of a star. Then by deducting 270° from the result its temperature in the ordinary Centigrade degrees is obtained. Thus in the case of the Sun the maximum heat radiation occurs in the greenish-yellow light, which gives a temperature for the rotating disk of the Sun of about 5,000° C., or 9,000° F. The atmospheric absorption already referred to serves to cut down the intensity of the radiation, so that, taking this and other amounts into consideration, the temperature of the Sun's disk can be estimated at about 6,200° C.

Similar investigation in the case of Sirius and Vega, which are white, or younger stars, give a temperature about 1,000° C. higher than that of the Sun, while the red star, Betelgeux, which is a declining star, older than the Sun, has a temperature some 2,500° C. less. The temperature of the Sun furnishes different results, depending on the manner in which the problem is attacked. Arrhenius, in his work on "Worlds in the Making," summarizes recent work, and states that: "From the intensity of the radiation, Christiansen, and afterward Warburg, calculated a temperature of about 6,000° C. Wilson and Gray found for the center of the Sun 6,200°, which they afterward corrected into 8,000°." Owing to the absorption of light by the terrestrial and the solar atmospheres, too low values are always found. That applies to a still greater extent to any estimate based upon the determination of that wave length for which the heat emission from the solar spectrum is maximum. Le Chatelier compared the intensity of sunlight filtered through red glass with the intensities of light from several terrestrial sources of fairly well-known temperatures treated in the same way. These estimates yielded to him a solar temperature of 7,600° C. Most scientists accept an absolute temperature of 6,500°, corresponding with about 6,200 C. That is what is known as the "effective temperature" of the Sun.

If the solar rays were not partially absorbed this temperature would correspond with that of the clouds of the photosphere. Since red light is little absorbed, comparatively, Le Chatelier's value of 7,600° C., and the almost equal value of Wilson and Gray of 8,000° C., should represent approximately the average temperature of the outer portions of the clouds of the photosphere. The higher temperature of the faculæ is evident from their greater light intensity, which, however, may be partly due to their greater height. Carrington and Hodgson saw on September 1, 1859, two faculæ break out from the edge of a sun-spot. Their splendor was five or six times greater than that of the surrounding parts of the photosphere. That would correspond with a temperature of about 10,000 or 12,000° C. The deeper parts of the Sun which broke out on these occasions evidently have a higher temperature, and this is not unnatural, since the Sun is losing heat by radiation from its outer portions.

At all events, the Earth receives a large amount of energy in the form of light and heat which amounts to three horsepower for every square yard of space perpendicular to the Sun's rays; and while this really is not available for mechanical purposes, yet solar engines have been constructed in which this energy has been transformed into power. While the energy of the Sun is not, generally speaking, available for mechanical purposes, yet indirectly the heat received by the Earth has made possible plant and animal life, on which depends the source of all energy.

The transmission of the Sun's heat to the Earth is one of the important problems of present-day physics and meteorology, inasmuch as the amount of radiation or heat emitted by the Sun, spoken of by physicists as the "solar constant" and the relation of this radiation to terrestrial temperature, as well as the study of the radiation of different parts of the Sun's disk, are all topics of fundamental importance. The solar constant is measured outside of the Earth's atmosphere at mean solar distance, and the intensity of radiation is employed for a unit, which, when fully absorbed for one minute over a square centimeter of area placed at right angles to the ray, would produce sufficient heat to raise the temperature of a gram of water 1° C.

If once the original quantity and kind of heat emitted by the Sun be known, its effect on the constituents of the atmosphere on its journey to the Earth, how much of it reaches the soil, how through the change of the atmosphere it maintains the surface temperature of our globe, and finally, how in diminished quantity, or altered kind, it is returned to outer space, it would be possible to predict nearly all of the phenomena of the weather. Thus it has been known that when there is a small decrease in the solar radiation there follows a marked and general decline in temperature. So that, knowing the variation of radiation, it should be possible to predict changes in climate.

These data are secured by measuring the total intensity of the radiation, as it arrives at the Earth's surface, with the pyrheliometer, or thermometer with blackened bulb, carefully protected from all other influences except the direct rays of the Sun; and in the second place by measuring the heat or energy in different parts of the solar spectrum with the spectro-bolometer. The absorption of the atmosphere nearer the Earth for different areas requires measurements to be made at several stations, for at Washington, near the sea-level, the intensity of radiation actually observed is only about three-fourths as great as that observed in the clear atmosphere of Mount Wilson, at a height of 6,000 feet. Recent observations made at Mount Wilson and Washington by Mr. Abbot, of the Astrophysical Laboratory of the Smithsonian Institution, indicate that heat sent out to the Earth from the Sun in the course of a year is capable of melting an ice shell 114 feet thick over the whole surface of the Earth. The solar radiation is not a constant quantity, but varies with the decrease in solar distance, the changes occurring from month to month and from year to year. The variation is due to the changes in the source of radiation rather than to the effects of our atmosphere or external causes.

The distance and position of the Sun as regards the other members of the Solar System has been considered in a previous chapter. It is known that its apparent diameter is 32' 4", which corresponds with 866,500 miles, or 109½ times the diameter of the Earth. This would give it an area 11,950 times that of the Earth and a volume 1,306,500 times greater. The actual mass of the Sun is 332,000 times that of the Earth, but its average density is only about a quarter as great, so that the Sun, which has a density of 1.41 as compared with water, is four times as large as it would have to be if its density were the same as that of the Earth. Taking into consideration this lightness as well as the high temperatures prevailing in the Sun, one is forced to the conclusion that the body of the Sun must be in a gaseous state. The conditions under which the gases are found must be quite different from those with which we are acquainted on the Earth. Gravity at the surface of the Sun exceeds by more than twenty-seven times gravity on the Earth.

The motion of the Sun as regards the other stars in the heavens has been treated elsewhere, but it is proper here to refer to its rotation on its axis from west to east, which takes place in a period of about 25 days and is apparent from the motion of the sun-spots, tho it can also be detected by directing a spectroscope toward the edges of the limb and noting by Doppler's principle how one side is approaching the observer and the other is retreating. The time of rotation is not the same for all parts of the disk, but depends upon the position of the spots selected. Those nearest the equator show the most rapid rotation.

CHAPTER XI

ANCIENT records make no mention of the discovery of Mercury, yet the existence of the planet was surely known even in the days of Nineveh, when a chief astronomer directly refers to the planet in a report which he made to King Assurbanipal of Assyria. The planet is occasionally mentioned in early and medieval astronomical literature. It is stated that Copernicus regretted that he had never been able to observe it properly in the high altitude of Frauenburg.

That the planet should have been overlooked by the ancients is not strange when it is considered that it is never visible in the higher altitudes except occasionally near the horizon, just after sunset or before sunrise. In the clear sky over an eastern desert the primeval astronomer doubtless saw the bright star in that part of the horizon where the setting Sun was still shedding its beams Its luster diminishes as the planet draws near the horizon at sunset, until finally it sets so soon after the Sun that it is invisible. Years may elapse before a similar opportunity is afforded.

If a similar phenomenon took place at sunrise, the primitive astronomer might have inferred that Venus and Mercury were identical, especially as a long series of observations would establish the fact that one of these bodies was never seen until the other had disappeared. Accordingly some of the ancient astronomers assumed that there was but one morning and evening star. But as records accumulated it was recognised that there were two bodies which might serve as morning and evening stars. A certain regularity in the recurrence of each planet was noted, and it was possible to make predictions of accuracy as to the time when either could be seen after sunset or before sunrise.

While by the time of Plato it was known that Venus and Mercury performed their revolution in approximately the same time, it was recognised that Mercury's period was different and more rapid than that of other planets. An older tradition, attributed to the Egyptians, stated that both planets revolved around the Sun. Ptolemy states that they could be regarded as oscillating to and fro on each side of the Sun. That the ancient astronomers might well have been confused by the appearance of Mercury as morning and evening star follows from a consideration of the planets' position and motion relatively to the Sun and the Earth. The consideration, moreover, applies to Venus as well.

Quoting C. G. Dolmage's "Astronomy of To-day," "when furthest from us Mercury is at the other side of the Sun, and cannot then be seen owing to the blaze of light. As it continues its journey it passes to the left of the Sun, and is then sufficiently away from the glare to be plainly seen. It next draws in again toward the Sun, and is once more lost to view in the blaze at the time of its passing nearest to us. Then it gradually comes out to view on the right hand, separates from the Sun up to a certain distance as before, and again recedes beyond the Sun, and is for the time being once more lost to view.

"To these various positions technical names are given. When the inferior planet is on the far side of the Sun from us it is said to be in 'Superior Conjunction.' When it has drawn as far as it can to the left hand, and is then as east as possible of the Sun, it is said to be at its 'Great-

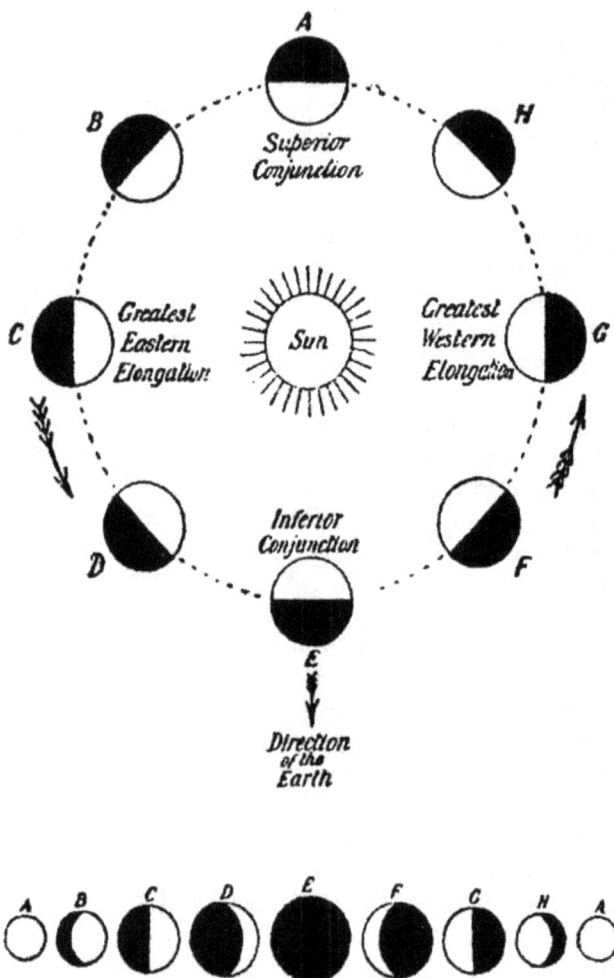

Fig. 18 —ORBIT AND PHASES OF AN INFERIOR PLANET.
Corresponding views of the same situations of an inferior planet
as seen from the earth, showing consequent phases and alterations in apparent size.

est Eastern Elongation.' Again when it is passing nearest
to us it is said to be in 'Inferior Conjunction'; and finally,
when it has drawn as far as it can to the right hand, it is
spoken of as being at its 'Greatest Western Elongation.' "

The continual variation in the distance of the inferior
planets, Venus and Mercury, from the Earth, during their
revolution around the Sun, will of course be productive of
great alterations in apparent size, which no doubt also
had its effect in bewildering the ancients. At superior
conjunction, being then farthest away, Mercury ought to
show the smallest disk; while at inferior conjunction, being the nearest, it should appear much larger. When at
greatest elongation, whether eastern or western, it should
naturally present an appearance midway in size between
the two.

From these considerations it would seem that the best
time for studying the surface of Venus or Mercury is at
inferior conjunction, or when the planet is nearest to the
Earth. But that this is not the case will at once appear
if it is considered that the Sun's light is then falling upon
the side most distant, leaving the proximate side unil-
lumined. In superior conjunction, on the other hand,
the light falls full upon the side of the planet facing the
Earth; but the disk is then so small and the view besides
is so dazzled by the proximity of the Sun that observations are of little avail. In the elongations, however, the
sunlight comes from the side, and so we see one-half of
the planet lit up; the right half at eastern elongation and
the left half at western elongation. Piecing together the
results given at these more favorable views, it is possible,
bit by bit, to gather some small knowledge concerning the
surface of Mercury or Venus.

Consequently it will be seen that the inferior
planets show various phases comparable with the wax-
ing and waning of the Moon in its monthly round.
Superior conjunction is, in fact, similar to full Moon,
and inferior conjunction to new Moon; while the eastern

and western elongations may be compared respectively with the Moon's first and last quarters. When these phases were first seen by the early telescopic observers, the Copernican theory was felt to be immeasurably strengthened; for it had been pointed out that if this system were correct, the planets Venus and Mercury, were it possible to see them more distinctly, would of necessity present phases when viewed from the Earth.

The apparent swing of an inferior planet from side to side of the Sun, at one time on the east side, then passing into and lost in the Sun's rays, to appear once more on the west side, is the explanation of what is meant when one speaks of an "evening star" or a "morning star." Mercury or Venus is called an "evening star" when it is at its eastern elongation—that is to say, on the left hand of the Sun; for, being then on the eastern side, it will set after the Sun sets, as both sink in their turn below the western horizon at the close of day. Similarly, when either planet is at its western elongation—that is to say, to the right hand of the Sun—it will precede him, and so will rise above the eastern horizon before the Sun, receiving therefore the designation of "morning star.

Mercury's motions were early studied. With the planet was associated the first of the remarkable astronomical predictions that now have become almost commonplace, so carefully are they worked out by astronomers. Kepler, from his studies of the motions of the planet, was the first to realize that if the orbit of Mercury, which, as has been seen, lies within that of the Earth and thus nearer the Sun, were exactly circular and in the plane of the ecliptic, a transit of Mercury across the Sun's disk would occur once in each synodical revolution, or period between two successive conjunctions of the planet with the Sun as seen from the Earth, the epoch of the transit could be calculated very easily. But Mercury's orbit is inclined 7 degrees to the ecliptic and is very eccentric, for which reason his calculations were only approximately correct.

It was in 1627 that Kepler predicted transits of both Mercury and Venus across the Sun, and assigned the date of November 7th for the former. Gassendi, to whom was entrusted the task of proving Kepler's prophecy, began his observation on the 5th of November, watching the image of the Sun formed by a lens on a white screen from light admitted through a hole in a dark room. But Kepler was not in error by as much as that. Only five hours after the time assigned did the transit actually begin. Thus was commenced a series of observations of the time when Mercury crosses the great central luminary of this system. These transits are by no means rare, for thirteen of them were observed in the nineteenth century. They afford opportunity for observation which enables the movements of the planet to be calculated with accuracy.

After the transits of 1661 and 1677 La Hire constructed new tables of Mercury and predicted a transit on May 5, 1707. The calculation was in error by nearly a day. After the transit of May, 1753, had been observed several hours later than La Hire's and as much earlier than Halley's prediction, Lalande constructed new tables, and predicted the transit of 1786, which actually took place fifty-three minutes later than the time announced by Lalande and about as much earlier than the time computed by Halley's tables. Lalande then corrected his tables and predicted the transits of 1789, 1799 and 1802 with a fair approach to accuracy. The tables compiled by Lindenau in 1813 still left much to be desired.

The problem of Mercury's motion was in this condition when it was attacked by Leverrier, who was under no illusion in regard to its difficulty. Leverrier did not succeed in overcoming the difficulty until 1859. As a result of his work, supplemented by that of later investigators, astronomers are now in possession of at least fairly accurate information on the subject of the planet's orbital phenomena. It is known, for example, that Mercury is the swiftest in its movements of all the planets. With the exception of some of the satellites, it has an orbit that departs most from the circular form—in other words, an orbit marked not only by the greatest eccentricity, but also by the greatest inclination of all planetary orbits to the ecliptic. This eccentricity of the orbit is such that the Sun is seven and one-half million miles out of its center, while the actual distance of the planet from the Sun ranges from about twenty-nine million miles to forty-three million, or a mean distance of about thirty-six millions of miles. The velocity varies from thirty-five miles a second, when the planet is at perihelion or nearest the Sun at the pointed end of its oval course, to about twenty-three miles a second at aphelion.

To explain the perturbations in Mercury's motion, Abbé Moreux has recently advanced the following hypothesis: "The Sun is unquestionably surrounded by matter which gives rise to the observed phenomena of the corona. Photographs made during the eclipse of 1905 showed an outer corona extending to a hitherto unsuspected distance from the Sun. This outer corona is ellipsoidal and its major axis is very nearly coincident with that of the zodiacal light. This region, then, is filled with a resisting medium, composed of matter which is gradually falling in toward the Sun. The mechanism of this fall, or contraction, is very complex, and as we have no idea of the density of the matter, it is difficult to calculate its changes in form.

"The resistance which the medium opposes to a moving body is likewise difficult to estimate, but we know from the observed acceleration of comets that this resistance is by no means negligible. At the epochs of maximum coronal development this resisting medium may easily extend to the orbit of Mercury. Hence that planet in its revolution around the Sun must traverse masses of rarefied gas and swarms of minute particles by which its course is modified. We are far from knowing the quantitative effect of the resisting medium, and much study of the corona will be required to complete the solution of this interesting problem."

Because of the lack of surface markings the rotation periods are very uncertain. Recently Professor McHarg, in France, has deduced a rotation of 24 h. 8 m. G. Schiaparelli has put forward the view that both Venus and Mercury rotate in a time equal to the individual period of revolution around the Sun, and thus always turn the same face toward the Sun. Such a motion, which is analogous to that of the Moon around the Earth, could be easily explained as the result of tidal action at some past time when the planets were, to a great extent, fluid. This may be true, as Mercury is one of the planets which has no Moon and consequently will be influenced directly and solely by the Sun in so far as tidal phenomena are concerned. These tidal effects must be especially severe on account of the proximity of Mercury to the Sun.

Because of this lack of a satellite the mass of the planet is very difficult to determine. At various times Laplace, Encke and Leverrier have given values from 1/9 to 1/30 of the Earth's mass. The first convenient opportunity of placing this planet in a gravitational balance and of determining its mass was presented in August, 1835, when Encke's comet in the course of its eccentric path through the solar system penetrated within the orbit of Mercury at its perihelion. The attractive power of the planet produced only a slight deviation from the regular course of the comet, sufficient, however, to show that the value previously assumed for Mercury's mass was nearly twice too large. At the same time the calculations made from these data have not removed the uncertainty that still prevails upon this point; Simon Newcomb decided on a value of about 1/21 that of the Earth, while William Harkness made it 1/25.

Beyond its motion comparatively little is known of Mercury. The spectroscope seems to indicate that the atmosphere of Mercury contains water vapor just as does the Earth. But the studies of Professor Lowell at Flagstaff, Arizona, do not show any signs of clouds or obscurations, and there are no indications of any atmospheric envelope. In fact, the surface of Mercury has been termed colorless, "a geography in black and white."

The first student of the surface of Mercury was Johann Hieronymus Schröter (1745-1816), who, as will be seen, was a pioneer in the study of the topography of the Moon. In April, 1792, Schröter concluded from the gradual degradation of light on its brightly illuminated disk that Mercury possessed a tolerably dense atmosphere. During the transit of May 7, 1799, he was struck with the appearance of a ring of softened luminosity circling the planet to a height of 3″ and about a quarter of its own diameter.

Referring to this ring of softened luminosity, Agnes Clerke writes in her 'History of Astronomy': "Altho a 'mere thought' in texture, it remained persistently visible both with the seven-foot and the thirteen-foot reflectors, armed with powers up to 288. A similar appendage had been noticed by De Plantade at Montpellier, November 11, 1736, and again in 1786 and 1789 by Prosperin and Flaugergues; but Herschel, on November 9, 1802, saw the preceding limb of the planet projected on the Sun cut the luminous solar clouds with the most perfect sharpness. The presence, however, of a 'halo' was unmistakable in 1832, when Professor Moll, of Utrecht, described it as a 'nebulous ring of a darker tinge approaching to the violet color.' Again to Sir William Huggins and Stone, on November 5, 1868, it showed as lucid and most distinct.

"No change in the color of the glasses used, or the powers applied, could get rid of it, and it lasted throughout the transit. It was next seen by Christie and Dunkin at Greenwich, May 6, 1878, and with much precision of detail by Trouvelot at Cambridge, Mass. Professor Holden, on the other hand, noted at Hastings-on-Hudson the total absence of all anomalous appearances. Nor could any vestige of them be seen by Barnard at Lick on November 10, 1894. Various effects of irradiation and diffraction were, however, observed by Lowell and W. H. Pickering at Flagstaff; and Davidson was favored at San Francisco with glimpses of the historic aureola, as well as of a central whitish spot which often accompanies it. That both are somehow of optical production can scarcely be doubted."

The planet's physical condition presents many problems and the spectroscope affords little information as to its constitution. As it shines only by reflected light from the Sun, its spectrum is but a fine echo of the Fraunhofer lines. An atmosphere like that of the Earth was suspected, however, by H. C. Vogel in 1871 from his spectroscopic studies, tho on the very slightest grounds.

Later observations made at the Potsdam Observatory by Müller confirm this conclusion and demonstrate that Mercury has a rough rind of dusky rock, which absorbs all but 17 per cent. of the direct light radiated upon it by the near-by Sun. This would seem to show the utter absence of any appreciable Mercurian atmosphere. The probable lack of atmosphere is also shown by the circumstance that when Mercury is just about to transit the face of the Sun, no ring of diffused light is seen to encircle its disk, as would be the case if it possessed an atmosphere.

According to the Belgian astronomer, Stroobant, the linear diameter of Mercury expressed as a fraction of the Earth's is 0.350 instead of the previously accepted 0.373, the radius 2,232 kilometers (1,386.7 miles), and the volume, compared with that of the Earth, 0.042.

CHAPTER XII

For all time Venus has been known as evening and morning star. By the ancients it was supposed to be two separate bodies which were named Phosphorus (Lucifer) or Morning Star and Hesperus (Vesper) or Evening Star. For Pythagoras in the sixth century B.C. is claimed the honor of discovering the identity of the two stars, but it was probable that he restated the views of Eastern astronomers.

As the most brilliant star of the heavens, Venus is certainly the one that has always been most observed. Glittering in the sky like a clear diamond, its pure white light, which, on a night when there is no Moon, is strong enough to cast a shadow, naturally would impress all whose lives are spent out of doors. Hence Venus has always figured in literature as the "Shepherds' Star." Homer speaks of the planet as Kallistos—the "Beautiful," and as a type of beauty its worship figures in all mythologies. At a time when the planets were personified as gods and goddesses it is easy to understand why this star was selected to typify Love.

Not only on account of its splendor in the evening sky, but for other phenomena, Venus is familiar in history and literature. The historian Varro (116-28 B.C.) states that "Æneas on his voyage from Troy to Italy saw this planet constantly above the horizon," and the same historian is quoted by St. Augustine as speaking of a change in the

color and brilliancy of Venus. In 1716 the visibility of
Venus in full daylight was hailed as a marvel by the people of London, and in 1750 its appearance at noon aroused
general astonishment in Paris. Again, in 1797, when
Napoleon returned to Paris from his conquest of Italy he
found the attention of the populace divided between his
reappearance and a similar striking midday phenomenon.
In fact, Bonaparte always associated this star with his
fortunes, and one evening while engaged in viewing the
starry heavens, pointing to the planet, said to Prince Tal-
leyrand, "Do you see? That is my star! So long as it
shines I will have no doubt of success."

In more recent times many brilliant appearances of the
planet have been observed, the interval between periods
of greatest visibility amounting to eight years. At one
of these, in April, 1905, at Cherbourg, Venus appeared
as a bright meteor of appreciable size and full form, an
effect which was due partly to the formation of a halo.

Acording to the Ptolemaic theory, Venus was always
between the Earth and the supposed orbit of the Sun.
Hence it was possible that, at the most, but half of its
illuminated surface could be visible to the Earth. When
Galileo, with his telescope, in 1610, made the striking
discovery that Venus appeared in various phases just as
the Moon, exhibiting the gibbous as well as the crescent
phase, it was a strong and almost the last argument necessary to establish beyond question the Copernican theory
that the planets revolve around the Sun. Were Venus
as large as Jupiter, for example, the phases would readily be discerned by the unaided eye. Indeed, Sir Robert
Ball has wondered what would have been the effect on the
history of astronomy had Venus been of the size of Jupiter so that its crescent form could have been seen without
a telescope. Then the elementary truth would have been
apparent that Venus was a dark body revolving around the
Sun. The analogy between it and our Earth would have

been at once perceived and the theory of Copernicus long since might have been established.

The mean distance of Venus from the Sun is 67,200,000 miles. With the exception of the Moon and an occasional comet no other heavenly body comes so near the Earth at any time. The orbit is marked by the smallest eccentricity in the planetary system, and is therefore more nearly circular than that of any other planet. The planet's great-

Fig. 19 —THE FOUR PRINCIPAL PHASES OF VENUS.

est and least distances from the Sun vary from the mean only by about 470,000 miles each way.

Venus is the nearest planet to the Earth, as there is only 26,000,000 miles between the orbits of the two at inferior conjunction or their nearest approach, yet Venus is not as well known as Mars, since when Venus passes nearest the Earth it is then between it and the Sun, so that the hemisphere which is illuminated is not visible to the Earth.

The appearance of Venus in the sky as the evening and the morning star was no less impressive to the ancients

than the beautiful character of the star itself. It is as
familiar now as it was to the shepherds of old that when
Venus disengages itself in the evening from the rays of
the setting Sun it departs from the Sun a little more
every night, increasing its brilliancy until a certain distance in the east is reached, appearing, like the Moon, to
travel toward the left of the observer. At the end of a
few months it has removed itself from the Sun to an
angular distance that may amount to as much as 48°, at
which time the planet sets more than three hours after
the Sun. After shining for some months, little by little the
planet begins to return toward the Sun, receding more
and more from the Earth, then passing behind the central
luminary and thus ceasing to be the evening star.

After an interval a new star is seen in the early morning to precede the rising of the Sun, advancing by imperceptible degrees every day, and eclipsing likewise all the
bodies of the heavens by its dazzling light. At this time it
proceeds toward the west, that is, toward the observer's
right hand, and we have now the "morning star." After
having preceded the rising of the Sun by three hours,
Venus resumes its course anew toward the Sun and
again is lost in the glare of day. It is then passing between the Sun and the Earth and is at its greatest
proximity to the Earth. Sometimes it passes just in front
of the Sun, as it did in 1874 and 1882, which phenomenon
is known as a transit. As it happens but twice in a century a transit is an occurrence of considerable importance
to astronomers.

These transits are valuable for the easy determination
of the position of the planet, for the investigation of its
atmosphere and for the determination of the solar parallax by comparing the amount of apparent displacement
in the planet's path across the solar disk when the transit
is observed at widely separated stations on the Earth's
surface. These transits occur in June and December, taking
place in cycles whose intervals are 8, 105.5, 8, 121.5 years.

They have occurred on the following dates: Dec. 7, 1631; Dec. 24, 1639; June 5, 1761; June 3, 1769; Dec. 9, 1874; Dec. 6, 1882; and will occur again on June 8, 2004, and June 6, 2012.

The first observed transit, namely, that of 1639, was watched in England by two persons, Jeremiah Horrocks and a friend, William Crabtree, whom Horrocks had fore-warned of its occurrence. That the transit was observed at all was due entirely to the remarkable ability of Horrocks. According to the calculations of Kepler, no transit

Fig. 20 —THE TRANSIT OF VENUS. PATH ACROSS THE SUN IN THE TRANSITS OF 1874 AND 1882.

could take place that year (1639), as the planet would just pass clear of the lower edge of the Sun. Horrocks, however, worked the question out for himself, and came to the conclusion that the planet would actually traverse the lower portion of the Sun's disk. The event proved him to be right. Horrocks, who was said to have been a veritable prodigy of astronomical skill, unfortunately died about two years after this celebrated transit, in his twenty-second year.

The transits of Venus next observed in 1761 and 1769 were taken advantage of by Edmund Halley to suggest a means of ascertaining the distance of the Sun. The idea had originated in rather vague form with Kepler, but was suggested more definitely by James Gregory in 1663. After Halley had observed the transits of Mercury in 1677 he realized the advantages of the method and published several papers urging preparations for observing the transit.

"He pointed out," says Berry, treating of this point in his "Short History of Astronomy," "that the desired result could be deduced from a comparison of the durations of the transit of Venus as seen from different stations of the Earth, i.e., of the intervals between the first appearance of Venus on the Sun's disk and the final disappearance, as seen at two or more different stations. He estimated, moreover, that this interval of time, which would be several hours in length, could be measured with an error of only about two seconds, and that in consequence the method might be relied upon to give the distance of the Sun to about 1/500 part of its true value. As the current estimates of the Sun's distance differed among one another by 20 or 30 per cent., the new method, expounded with Halley's customary lucidity and enthusiasm, not unnaturally stimulated astronomers to take great trouble to carry out Halley's recommendations. The results, however, were by no means equal to Halley's expectations."

Immense trouble was taken by governments, academies

and private persons in arranging for the observation of the transits of 1761 and 1769. For the former, observing parties were sent as far as to Tobolsk, St. Helena, the Cape of Good Hope and India, while observations were also made by astronomers at Greenwich, Paris, Vienna, Upsala and elsewhere in Europe. The next following transit was observed on an even larger scale, the stations selected ranging from Siberia to California, from the Varanger Fjord to Otaheite (where no less famous a person than Captain Cook was placed), and from Hudson's Bay to Madras.

The expeditions organized on this occasion by the American Philosophical Society may be regarded as the first of the contributions made by America to the science which has since owed so much to her; while the Empress Catherine of Russia bore witness to the newly acquired civilization of her country by establishing a number of observing stations on the soil of her Empire.

A variety of causes prevented the moments of contact between the discs of Venus and the Sun from being observed with the precision that had been hoped. The values of the parallax of the Sun deduced from the earlier of the two transits ranged between 8" and 10"; while those obtained in 1769, tho much more consistent, still varied between 8" and 9", corresponding with a variation of about 10,000,000 miles in the distance of the Sun. The whole set of observations was subsequently very elaborately discussed in 1822-4 and again in 1835 by Johann Franz Encke, who deduced a parallax of 8".571, corresponding with a distance of 95,370,000 miles, a number which long remained classical. The uncertainty of the data is shown by the fact that other equally competent astronomers have deduced from the observations of 1769 parallaxes of 8".8 and 8".9.

The transits of Venus in 1874 and 1882 were memorable as the first in which photographic methods and the heliometer were applied, in addition to the older methods of time observation to measure the occasion of contact. The older methods left unavailable the remainder of the transit, but the heliometer made it possible to ascertain the planet's apparent position upon the Sun's disk at any time with great precision. In the transit of 1874 the most elaborate photographic measures ever undertaken until this time, were carried out in the United States where conditions were most favorable.

A continuous succession of photographs was made as the planet traveled across the Sun's disk, and various systems of reference enabled its position to be identified with great accuracy at any part of the transit. Yet results were hardly more valuable than those obtained by the other methods. Hence in 1882 the American observers modified their apparatus considerably from previous practice by employing the photographic plate and using the heliostat to reflect the Sun's rays through a telescope lens 5 inches in diameter and of 40-foot focal length. Doubtless better results could be obtained to-day with modern solar photographic equipment and dry plates. No great advances were recorded as the result of the numerous expeditions during these two transits.

Elaborate expeditions were equipped, particularly for the second of the two transits, and the observations, which were voluminous, were thoroly and systematically discussed. The results were a somewhat larger value of the solar parallax, amounting to 8″.857, or rather less than 93,000,000 miles, as obtained by Newcomb after combining the heliometer with photographic measurements. As all measures of this kind seem to be affected by some constant systematic error, it is the concensus of opinion among astronomers that other methods are more to be trusted in determining the solar parallax than those based on the transits of Venus.

The great brilliancy of Venus is due to its relatively small distance from the Sun and the Earth and to its great reflecting power. The albedo of Venus, or the ratio of reflected to incident light, is 0.76. In other words, the planet reflects three-fourths of the light that falls upon it from the Sun, a proportion which is little exceeded by freshly fallen snow. This indicates that Venus is surrounded by a dense and cloudy atmosphere. Similar indications are given by the spectroscope, pointing to the existence of an atmosphere containing water vapor and generally similar to our own, but denser.

This atmosphere of Venus, owing to its great refractive power, sometimes appears as a luminous ring in transits of the planet. From observations made at the transit of 1882 it has been computed that the density of the atmosphere of Venus is 1.8 that of the Earth's atmosphere, which produces a refractive displacement of 33″. According to Maedler's calculations, the atmosphere of Venus is 1.7 times as dense as that of the Earth.

Many astronomers believe therefore that Venus is surrounded by a dense atmosphere into which the Sun's rays do not penetrate to a depth sufficient to cause much loss by absorption—probably because they are reflected by opaque clouds. From this it would follow that the markings which have been observed on the disc of Venus through the telescope for centuries may be purely atmospheric. As the displacement of the surface markings of a planet when viewed through a telescope furnish the basis for estimates of the period of rotation, it may be that calculations so made are without justification.

These markings of Venus are altogether different from those of any other planet. They are faint, indistinct and apparently variable, for the drawings of Venus made by different observers before the advent of photography are very unlike. The first astronomer to observe any markings on Venus was Domenico Cassini, who discovered a bright spot in 1666. In the following year he discovered a second spot, from which he deduced a period of rotation of about 24 hours, an estimate which was reduced to 23 hours and 22 minutes in Jacques Cassini's revision of his father's work. On the other hand, Bianchini, who studied the planet in 1726-1727, deduced from his observations a period of rotation equal to 24 days and 8 hours.

Thus began the long and violent controversy in regard to the rotation of Venus. From more than 10,000 observations made between 1830 and 1841 De Vico computed a mean value of 23 hours, 21 minutes and 23.93 seconds for the period of rotation of Venus. This value, furthermore, agrees very closely with the period of 23 hours and 21 minutes deduced by Schröter from observations of the deformation of the horns of Venus (1788-1793), and it has consequently been adopted in most treatises on astronomy. But in 1890, however, belief in the short period was shaken by Schiaparelli's discoveries. Before this Schiaparelli had made the surprising announcement that the period of rotation of Mercury was equal to the period of its revolution about the Sun, and he now asserted that the same law governed the motions of Venus, a statement based on observations made in the winter of 1877-78, which showed that the bright spots then visible never varied their positions with respect to the terminator, or boundary line between the planet and its shadow. Observations made in 1895 gave additional support to the view that Venus rotates on her axis in the period of her revolution about the Sun (225 days), and consequently always turns the same hemisphere toward the Sun, just as the Moon always presents the same face to the Earth.

This theory, which on its announcement rested solely on Schiaparelli's authority, soon obtained strong independent support. It has apparently been completely confirmed by the observations made by Lowell at Flagstaff, Arizona, in 1895-6. Lowell saw markings that have been seen by no other astronomer, before or since, but the most surprising thing about his discoveries is that he saw no spots, but only bands or lines bearing a superficial resemblance to the canals of Mars. The whole configuration remained unchanged for hours, which could not be the case if the period of rotation were only 24 hours. Furthermore, the markings remained fixed relatively to the terminator. Lowell denies the existence of clouds on Venus, altho he finds in certain phenomena of twilight evidence of the presence of a very dense atmosphere.

Many astronomers have refused to accept the conclusions of Schiaparelli, Perrotin and Lowell, and have adhered to the theory of a short period of rotation. Schiaparelli's first announcement was vigorously disputed, soon after its publication, by the French astronomer, Trouvelot. During his residence at Cambridge, Mass., from 1875 to 1882, Trouvelot pointed out seventeen changes in the markings which Schiaparelli regarded as fixed. Trouvelot regarded the bright spots on the limb of the planet as very high mountains, rising above the atmosphere.

The observed changes in the appearance of the horns of the crescent planet have also been ascribed to mountains. In the eighteenth century Schröter announced the discovery of several mountains on Venus. In 1789 the southern horn of the crescent appeared blunted and near it on the dark disc of the planet shone a bright peak, the height of which was estimated by Schröter as 117,000 feet, or 22 miles. Trouvelot deduced from his own observations a period of rotation of about 24 hours. It is very remarkable that two such experienced and trustworthy observers as Schiaparelli and Trouvelot could be led to such widely differing results from their frequent observations of the same object during the same period. It may be regarded as certain that the problem of the rotation of Venus cannot be solved by observing markings, which is not surprising if they are purely atmospheric phenomena.

Sir William Herschel, who has justly been called the greatest observer of all time, doubted the reality of the markings of Venus and regarded them as atmospheric phenomena, but this view was not shared by any of his contemporaries. Beer and Maedler, who were also most excellent observers, studied Venus repeatedly in 1833 and later years, but they rarely, if ever, saw anything that deserved the name of a marking.

Attempts have been made to solve the interesting and difficult problem of the rotation of Venus with the aid of the spectroscope, but these attempts have also failed to give a positive result. The effort was first made by the Russian spectroscopist, Bjelopolsky, at the Pulkowa observatory. The spectrophotographic measurements which were begun in 1900 indicated a short period of rotation. Although Bjelopolsky did not regard the correctness of the result as assured, his authority led to the belief that the spectroscope had decided the question in favor of the short period. Then Lowell, who had deduced a period of 225 days from telescopic observations, sought confirmation for this value by the spectroscopic method. He first tested his new and delicate apparatus and method by applying them to Mars, and obtained for that planet a period of rotation of 25 hours and 35 minutes, which agrees fairly well with the known period, 24 hours and 37 minutes. For Venus, Lowell's spectrophotographs give a rotational velocity of from 5 to 8 meters (16.4 to 26.2 ft.) per second, which corresponds with a long period of rotation, tho one much shorter than 225 days, for which the equatorial velocity would be about 2 meters (6.56 ft.) per second.

Neither of the results obtained at Pulkowa and at Flagstaff has yet been confirmed elsewhere. The spectroscopic method is too difficult to be attempted except by a particularly well equipped and favorably situated observatory. The theory of slow rotation receives powerful support from certain arguments advanced by the distinguished English astronomer, Sir George H. Darwin. The rotational velocity of every planet must be gradually diminished by tidal friction among its parts. In this way the period of the Moon's rotation has been made equal to the period of her rotation about the Earth, and the rotational periods of Mercury and Venus may have been similarly lengthened to equality with their periods of revolution about the Sun, which exerts a far greater tidal action upon these inferior planets than upon the Earth.

If Venus has already reached this stage, all the water on the planet must be accumulated in the form of ice on the dark and intensely cold hemisphere. Hence, as Antoniadi has pointed out, the presence of clouds and water vapor on the illuminated hemisphere casts grave doubt on the correctness of Schiaparelli's theory. The presence of clouds in the atmosphere of Venus is denied only by Lowell, who saw the peculiar canal-like markings distinctly whenever terrestrial atmospheric conditions permitted.

No trace of markings, by the way, was seen by Hansky and Stefanik, who studied the planet in the summer of 1907 at the Mont Blanc observatory, which is more elevated than Flagstaff, under very favorable atmospheric conditions, and deduced a period of rotation slightly less than 24 hours. So the problem of the rotation of Venus is still unsolved, altho a majority of astronomers appear to share the view of Schiaparelli and Lowell. The hope that it can be solved by observing markings has proven illusory, but there is good reason to believe that the spectroscope will ultimately furnish the solution.

Venus in size is almost identical with the Earth. Its diameter expressed in miles is 7,800, as compared with 7,920 for the Earth. Its circumference consequently amounts to 23,400 miles, as compared with 25,000 miles, that of the Earth. The period of revolution around the Sun is 225 days.

For many years observers have been looking for satellites of Venus. Fontana (1645), Cassini (1672 and 1686) and Montaigne (1761), among others, imagined that they had made such a discovery, but these have for many years been considered optical illusions.

CHAPTER XIII

THE EARTH

THE conception of the Earth formulated by the ancients assumed it to be a flat disk floating on water, which idea persisted in some form or other despite various phenomena clearly impossible of explanation upon such a hypothesis. After some more or less vague speculation on the shape of the Earth by Thales (640[?]-546 B.C.[?]), a clear and rational idea was advanced by Pythagoras, who flourished in the sixth century B.C. and taught that the Earth, in common with the heavenly bodies, was a sphere and that it received no support at the center of the universe. The idea of the sphericity of the Earth was doubtless derived by analogy from the appearance of the Moon. The theory became an established part of Greek science and philosophy and consequently has continued to the present day, anticipating by some 2,000 years the acceptance of the belief in the Earth's rotation. In fact, the spherical form of the Earth figured prominently in Greek astronomy, and the division of the Earth into zones and the idea of poles was promulgated by the Alexandrian philosophers and even before their time. Indeed, not only did they realize the shape of the Earth, but also its size. Eratosthenes (276-195 or 196 B.C.), whom we have already mentioned in our chapter on pretelescopic work, made one of the first scientific measurements of the Earth, obtaining a result which must be considered the first good geodetic measurement. He found from the application of simple geometry that the angular distance of the Earth's surface between Alexandria and Syene must be $\frac{1}{50}$ of the circumference of the Earth. Posidonius, who was born about the end of the life of Hipparchus, made a new measurement of the circumference of the Earth, much after the fashion of that of Erathosthenes, but reaching a value too small. One of the crucial arguments for the sphericity of the Earth is that when a ship sails away the hull first disappears from view while the masts are visible. This was first advanced by Pliny (23-79 A.D.).

From the Greeks and Romans to the Arabs may be a far journey, but little was done in the study of the Earth until the Arab astronomers at Bagdad, under the patronage of Caliph Al Mamun, made a series of measurements of a meridian of the Earth which agreed with those of Ptolemy. These measurements virtually sufficed until Willebrod Snell (1591-1626), a Dutch mathematician who had discovered the law of the refraction of light, made a series of measurements of the Earth's surface from which he computed the length of a degree of a meridian to be about 67 miles, an estimate subsequently corrected to about 69 miles by one of his pupils and differing but a few hundred feet from the value now accepted. A measurement by Richard Norwood (1590[?]-1675) of the distance from London to York, made in 1636, enabled a degree of latitude to be obtained with an error of less than half a mile, and finally Picard, in 1671, made a measurement of this quantity, which we have seen was sufficiently accurate to be used by Newton in computing the mass and motion of the Earth and in demonstrating his law of gravitation.

This outline shows that the spherical shape of the Earth was constantly held in mind by scientists for about 2,500 years. Other views of the Earth's shape also obtained. The idea of a flat Earth had to be vigorously opposed by arguments founded on experiment, observation and discovery. After the phenomenon of a disappearing ship, one of the most important of the arguments against the flatness of the Earth's surface was based on the different elevations of the Pole Star in the sky, depending upon the position of the observer. Now the Pole Star and the surrounding stars, tho they may be at different heights in the sky, form the same pattern wherever on the Earth's surface they are seen, thus showing that they are immensely distant and that lines from all points of the Earth's surface directed to the Pole Star would be parallel or have the same direction. As one travels north it is seen that the Pole Star sinks lower and lower in the heavens or has a greater angular distance from our zenith. If the Earth were flat this would not be the case, for an angular distance of 7½ degrees difference in the position of two stations 520 miles apart in a north-and-south line would mean that the star is only 3,450 miles distant, which is manifestly impossible. Furthermore, it is a well-known experience that in crossing the Atlantic from England to the United States the Sun, by a watch, rises an hour later for every 600 miles we travel west. If the Earth were flat the Sun would rise at the same instant over its entire surface. To answer both these conditions—namely, the sinking in the heavens of the Pole Star and the equal delays in the time of sunrise in traveling equal distances—a round globe is required.

Still the globular form of the Earth was not accepted generally until after the historic voyage of Christopher Columbus and that epoch of exploration and discovery which resulted in the circumnavigation of the Earth in 1519. After the demonstration of the sphericity of the Earth came the determination of its exact figure by the systematic measuring of arcs of meridians. When the results were mathematically examined the conclusion was reached that the Earth was not a perfect sphere, but flattened at the ends, or in other words that it was an oblate spheroid. Newton in his Principia, published in 1687, showed from theoretical considerations that the Earth bulged out at the equatorial regions. On the other hand, in the following century (in 1745) the Cassinis, who were leading astronomers of France and deeply interested in the measurement of the Earth's surface, maintained that the Earth was contracted at the equator and bulged at the poles, or in other words was prolate, the difference in the two ideas being concisely expressed by Professor Moulton, who likens Newton's idea of the Earth's figure to an orange and that of the Cassinis to a lemon.

It is interesting that Newton's proof of the oblateness of the Earth anticipated the direct observations of the astronomer and geodesist. To-day it can be demonstrated in simple form by means of observations of the time of vibration of a pendulum, such as are carried on systematically by such scientific agencies as the United States Coast and Geodetic Survey.

It is possible to supply various mathematical proofs of the oblateness of the Earth in connection with its motions and with its relations to the Moon and to the solar system. But these hardly concern us in the present pages, and we may dismiss the matter with the assertion that the ellipticity of the Earth as determined by Harkness in 1891 amounts to $\frac{1}{300}$, as compared with a value of $\frac{1}{295}$ established in 1866 by Colonel Clarke, of the English Ordnance Survey, and accepted for many years as a standard, and $\frac{1}{230}$, as was given by Newton for the ellipticity of a meridian section of this oblate spheroid. Clarke's spheroid of 1866, which is generally adopted in geodetic and other computations, would give a radius at the equator of 3,963.307 miles and 3,949.871 miles at the poles, these figures having been slightly corrected in 1878 in the second place of decimals. The radius of the Earth is in many ways a fundamental measure, especially in astronomy, and for that reason its determination is a matter of no small importance. Thus observations made with a view of finding the distance of the Moon, when discussed and reduced, simply give us this distance in terms of the equatorial radius of the Earth, so that to determine the distance in miles we must know the number of miles in the Earth's radius.

As a consequence of the spheroidal shape of the Earth a degree of latitude, as well as of longitude, varies with the point at which it is measured. Thus at the equator one degree of latitude is equal to 68.704 miles, but at the pole a similar degree is equal to 69.407 miles. The difference in a degree measured on the parallels of latitude, which are small circles, is of course obvious, and while at the equator a degree of longitude amounts to 69.652 miles, at latitude 40°, or approximately that of New York (40° 43'), it would equal 53.431 miles and at the north pole it would decrease, of course, to zero.

Newton in his researches on gravitation, by comparing the attraction exerted by the Earth with that of the Sun and other bodies, was able to connect its mass with that of the planets and Sun. The problem of determining the mass of the whole Earth in terms of a given terrestrial body—that is, in tons, pounds or kilograms—was indeed a serious one, and it was first attacked in Great Britain by Nevil Maskelyne (1732-1811), who was Astronomer Royal. If we know the volume of the Earth and its density, referred to some standard as water, air or iron, it is of course easy to determine its mass. But the determination of its density is no less a problem than the determination of its mass directly. It is possible, of course, to compute the relative amount of water, but this gives only a small amount of the mere surface of the Earth. We can descend in mine shafts a mile or more and investigate the nature of the material, but this again affords no clue to the thousands of feet of materials below, so that an estimate of the average density based on such data cannot be considered at all trustworthy. Newton, taking into consideration the various elements of the Earth, estimated its density at between five and six times that of water. Mas-

kelyne, however, bearing in mind the phenomenon of the
deflection of the plumb-line observed by Bouguer and
La Condamine in their expedition to measure an arc of
latitude in Peru (a phenomenon caused by the attraction
of the mountain Chimborazo), selected the narrow ridge
of Schehallien in Perthshire and found that the attraction
of the mountain caused a deflection of about 12″ on each
side of the ridge. This attraction of the mountain as
compared with the attraction of the Earth on the plumb-
bob enabled a value for the density of the Earth of about
4½ times that of water to be obtained. It remained for
the famous Cavendish experiment, carried out by Henry
Cavendish (1731-1810), to substitute for the mountain a
pair of heavy balls. This gave a value for the density of
the Earth of 5½ times that of water, a value confirmed by
numerous repetitions of Cavendish's classical experiment.
Expressed in pounds, the mass of the Earth is a little
more than 13 billion billion, a figure that the human mind
utterly fails to grasp.

There must have been a time when the Earth contained
much more heat than at present. The Sun is not supply-
ing heat enough to make good the losses sustained by
the globe. These losses must have been taking place
over a period of countless years. One is forced to believe
that not only once was our Earth much hotter than at present, but that its temperature was a red heat. Going back
still further, the now solid globe must have been actually
a molten mass. Being a molten mass, it is not difficult to
realize that it readily assumed a globular form on its journey through space. A falling drop of rain is a globe; a
drop of oil suspended in a liquid with which it does not
mix has a spherical shape. Therefore if a mass of material
as large as the Earth were so soft that its individual particles obeyed the forces of attraction exerted by each part
of the mass on all the other parts, it is quite obvious that
it would assume a globular shape, especially under the
influence of motion. If this great sphere were caused to

rotate around an axis, like a ball of clay on a potter's wheel, any lack of true sphericity would not only be overcome, but there would be a tendency for the molten body to bulge at the equator and flatten down at the poles, forming an oblate spheroid such as by actual measurement the Earth is found to be. Indeed, not only is the Earth an oblate spheroid, but most of the planets which can be measured have become flattened at the poles for the same reason.

In connection with the spheroidal shape of the Earth, mention should be made of a recent theory which seeks to demonstrate that the shape of the Earth is being transformed into a polyhedral form, or more exactly, that there is a tendency in the direction of its assuming the figure of a regular tetrahedron, or four-sided figure, each of whose faces is triangular. This is manifested by the lithosphere or solid portion of the Earth. The water is massed on the faces of such a tetrahedron, the force of gravity acting most powerfully at the center of each of the four surfaces. The tetrahedron hypothesis assumes that a more or less solid crust was formed when the Earth was still perhaps in an approximately spheroidal form. As contraction occurred in the interior the external shell proved too large. Hence a different form has to be assumed, which, it is asserted, would be the tetrahedral. On each of the faces of such a tetrahedron one of the four oceans would lie, the Arctic Ocean being assumed as the fourth for the region about the North Pole. The continents would correspond with North and South America, Euro-Africa, Asia-Australia and Antarctica. This hypothesis is borne out in many respects by geological and other considerations. First advanced in the 70's of the nineteenth century by Lowthian Green, the hypothesis has been put forward in recent years with further developments and attempts at its proof, so that it is considered as a tenable and possible explanation, tho its absolute soundness has not been demonstrated.

Recognising and proving the spherical shape of the

Earth by no means explains the apparent motion of the Sun, Moon and stars as regards the planet. As the nature of the solar system was developed by Copernicus and Newton, people were compelled to recognise that the Earth revolved around the Sun, and, furthermore, that it rotated daily about its axis.

These motions are necessary to explain various phenomena. For example, when a body moves in a circle a force is needed to act on it, pulling toward the center proportionally to the square of the rate of spin and proportionally to the distance. Every pound of mass in the Moon would need a pull equal to about the weight of 3 ounces to keep it in a circle which it traveled around once in 24 hours. Every pound in the Sun would need a pull about three-quarters of a hundredweight. The nearest fixed star would need a pull on every pound, of about 9,000 tons. One cannot imagine that the Earth could exert such forces and so are obliged to think that the Earth is revolving and not the sky.

Final proof of the rotation of the Earth was furnished in the middle of the last century by Foucault. His experiments are constantly repeated in physical laboratories and experimental lectures. From the dome of the Pantheon in Paris he suspended a heavy ball by a fine cord and caused it to swing in a single plane. The path of the ball across the floor could be traced, and it was found that after it had been in vibration for some time it was swinging in a different direction from that in which it had started. The reason was that the floor had revolved under the ball with the rotation of the Earth.

The daily rotation of the Earth on its axis and its yearly rotation about the Sun supply us with our means of measuring time. According to Professor Poynting, "The Earth is a clock, the line to the Sun is the finger and the sky is the face. But the Sun is not a regular timekeeper. Our twenty-four-hour day is only the average between successive noons or times when the Sun is due south. If compared with a good clock, the Sun is in parts of the year too soon and in other parts too late, sometimes as much as a quarter of an hour. This cannot be due to change in the Earth's rate of spin, for to change a spin there must be either a force acting at one side of the center of gravity or a change of shape. The forces on the Earth exerted by the Sun and Moon act almost exactly through the center of gravity and so affect the rate of spin hardly at all. The Earth does not change its shape sufficiently to account for the variation in the solar day.

"The variation in the solar day is due partly to the inclination of the Earth's axis to the plane in which it moves around the Sun, partly to variation of the Earth's motion round the Sun at different times of the year.

"The fixed stars keep good time, getting round in about 4 minutes less than 24 hours. By them clocks are rated. Their day is the true time of one revolution of the Earth."

Among the movements of the Earth there is one that is so minute as to be of peculiar interest to the astronomer. It is a small displacement or wobbling of the axis of the Earth or theoretical pole. This pole shifts its place through a circle of some few yards in diameter in the course of a period of somewhat over a year, and as a result there is a variation of latitude on the Earth's surface. This was discovered first in 1884 and 1885 by Dr. Küstner at Berlin, and after observations made at various stations by members of the International Geodetic Association in 1888, it was found that this synchronous variation was occurring in a periodic manner at a number of stations. The general nature of the occurrence was defined by Prof. S. C. Chandler in 1891, who stated that the pole of the Earth might be supposed to describe a circle with a radius of 30 feet in a period of fourteen months. One reason assigned for this phenomenon is a change in the distribution of mass, due to the temporary occurrence of heavy falls of snow or rain limited to one continent, or to the transportation of great bodies of air and water from place to place by atmospheric or ocean currents, so that the globe is made slightly lopsided and temporarily to forsake its normal axis. Also it has been supposed that the Earth is not absolutely rigid. As a quasi-plastic mass it would yield to certain strains which tend to protract the time of circulation of the displaced pole. In fact, in all consideration of the various phenomena of gravitation and tides we have to recognise that in practice the astronomer of to-day is not dealing with the theoretically rigid body of Newton and the earlier mathematicians.

One of the important conditions of the rotation of the Earth is that it moves around an axis which is not vertical but which is inclined toward the plane of its orbit. Dolmage in his volume, "Astronomy of To-day," tells us that "If the axis of the Earth stood 'straight up,' so to speak, while the Earth revolved in its orbit, the Sun would plainly keep always on a level with the equator. This is equivalent to stating that, in such circumstances, a person at the equator would see it rise each morning exactly in the east, pass through the zenith—that is, the point directly overhead of him at midday—and set in the evening due in the west. As this would go on unchangingly at the equator every day throughout the year, it should be clear that, at any particular place upon the Earth, the Sun would in these conditions always be seen to move in an unvarying manner across the sky at a certain altitude, depending upon the latitude of the place. Thus the more north one went upon the Earth's surface the more southerly in the sky would the Sun's path lie, while at the north pole itself the Sun would always run round and round the horizon. Similarly the more south one went from the equator the more northerly would the path of the Sun lie, while at the south pole it would be seen to skirt the horizone in the same manner as at the north pole. The result of such an arrangement would be that each place upon the Earth would always have one unvarying climate, in which case there would not exist any of those beneficial changes of season to which we owe so much.

"The changes of season which we fortunately experience are due, however, to the fact that the Sun does not appear to move across the sky each day at one unvarying altitude, but is continually altering the position of its path, so that at one period of the year it passes across the sky low down, while at another it moves high up across the heavens and is above the horizon for a much longer time. Actually the Sun seems little by little to creep up the sky during one-half of the year, namely, from mid-winter to mid-summer, and then just as gradually to slip down it again during the other half, namely, from mid-summer to mid-winter. It will therefore be clear that every region of the Earth is much more thoroly warmed during one portion of the year than during another—*i.e.*, when the Sun's path is high in the heavens than when it is low down.

"Once more we find appearances exactly the contrary from the truth. The Earth is in this case the real cause of the deception, just as it was in the other cases. The Sun does not actually creep slowly up the sky and then slowly dip down it again, but owing to the Earth's axis being set aslant, different regions of the Earth's surface are presented to the Sun at different times. Thus in one portion of its orbit the northerly regions of the Earth are presented to the Sun and in the other portion the southerly. It follows, of course, from this that when it is summer in the northern hemisphere it is winter in the southern and vice versa.

"The fact that in consequence of this slant of the Earth's axis the Sun is for part of the year on the north side of the equator and part of the year on the south side leads to a very peculiar result. The path of the Moon around the Earth is nearly on the same plane with the Earth's path around the Sun. The Moon, therefore, always keeps to the same regions of the sky as the Sun. The slant of the Earth's axis thus regularly displaces the position of both the Sun and the Moon to the north and south sides of the equator respectively in the manner we have been describing. Were the Earth, however, a perfect sphere such change of position would not produce any effect.

"At present the north pole of the heavens is quite close to a bright star in the tail of the constellation of the Little Bear, which is consequently known as the pole star, but in early Greek times it was at least ten times as far away from this star as it is now. After some 12,000 years the pole will point to the constellation of Lyra, and Vega, the most brilliant star in that constellation, will then be considered as the pole star. This slow twisting of the Earth is technically known as precession, or the precession of the equinoxes.

"We have seen that the orbit of the Earth is an ellipse and that the Sun is situated at what is called the focus, a point not in the middle of the ellipse, but rather toward one of its ends. Therefore during the course of the year the distance of the Earth from the Sun varies. The Sun, in consequence of this, is about 3,000,000 miles nearer to us in our northern winter than it is in our northern summer, a statement which sounds somewhat paradoxical. This variation in distance, large as it appears in figures, can, however, not be productive of much alteration in the amount of solar heat which we receive, for during the first week in January, when the distance is least, the Sun only looks about one-eighteenth broader than at the commencement of July, when the distance is greatest. The great disparity in temperature between winter and summer depends, as we have seen, upon causes of quite another kind, and varies between such wide limits that the effects of this slight alteration in the distance of the Sun from the Earth may be neglected for practical purposes."

The most striking effect of gravitation in its universal aspect is seen on the Earth in the case of the tides, which are due to the attraction of the Moon on the Earth and especially on the waters of its oceans. While the ancients could not explain the cause of the daily change in the tides, the phenomenon was obvious to all mariners or those dwelling by the sea or even on great rivers. The connection of the Moon with this phenomenon, it is claimed, was understood as early as 1000 B.C. by the Chinese, for at full and new moon exceptionally high tides were observed, and the name of spring tides was applied as contrasted with the minimum high water which was observed at the neap tides at the second and fourth quarters of the Moon. Not only the Chinese but the Greeks and Romans noticed this same phenomenon, and Julius Cæsar in his "Commentaries" tells us that when he was embarking his troops for Britain the tide was high because the Moon was full, while both Pliny and Aristotle connect the time of high water with the age of the Moon. The connection between the Moon and the tide was thus early established and practically applied by navigators, who were quick to realize the nature of the phenomenon, even tho they and the astronomers were unable to present any satisfactory explanation.

A general explanation of the tides as due to the disturbing action of the Moon and Sun, especially the former, was put forward by Newton. Newton's explanation, as described by Berry in his 'Short History of Astronomy,' was as follows: "If the Earth be regarded as made of a solid spherical nucleus, covered by the ocean, then the Moon attracts different parts unequally, and in particular the attraction, measured by the acceleration produced on the water nearest to the Moon, is greater than that on the solid Earth and that on the water farthest from the Moon is less. Consequently the water moves on the surface of the Earth, the general character of the motion being the same as if the portion of the ocean on the side toward the Moon were attracted and that on the opposite side repelled. Owing to the rotation of the Earth and the Moon's motion, the Moon returns to nearly the same position with respect to any place on the Earth in a period which exceeds a day by (on the average) about 50 minutes, and consequently Newton's argument showed that low tides (or high tides), due to the Moon, would follow one another at any given place at intervals equal to about half this period; or, in other words, that two tides would in general occur daily, but that on each day any particular phase of the tides would occur on the average about 50 minutes later than on the preceding day, a result agreeing with observation. Similar but smaller tides were shown by the same argument to arise from the action of the Sun and the actual tide to be due to the combination of the two. It was shown that at new and full moon the lunar and solar tides would be added together, whereas at the half moon they would tend to counteract one another, so that the observed fact of greater tides every fortnight received an explanation. A number of other peculiarities of the tides were also shown to result from the same principles." .

From Newton's time to the present the tides have supplied material for astronomical and mathematical research, so that through the efforts of many astronomers and other scientists their theory and occurrence are now well understood, and the various governments make ample provision through hydrographic offices and otherwise to provide such information for mariners. In the theoretical discussion in modern times one name stands out prominently, that of Sir George H. Darwin, of England, whose researches on the tides of the Earth have developed to a point where they have a most important bearing on cosmical theory.

It is desirable here to outline the chief features of tides in their everyday occurrence. The term "flood-tide" is applied to the rising water, which reaches "high-water" mark at the moment when the tide is highest, while in falling the tide is said to "ebb" until low water, or the lowest mark, is reached. Near the times of new and full moon occur the "spring tides," which are the highest tides of the month as distinguished from the "neap-tides," which are the smallest, occurring when the Moon is in quadrature, the relative heights of spring and neap-tides being about as 7 to 4. At the time of "spring-tides" the interval between the corresponding tides of successive days is less than the average value of 24 hours and 51 minutes and the tides are said to "prime," the interval amounting to about 24 hours and 38 minutes, while at "neap-tides" the interval increases to 25 hours and 6 minutes, and then the tides are said to "lag." The highest tides of all occur when the new or full moon occurs at the time the Moon is in perigee, especially as this takes place about January 1st, when the Earth is nearest to the Sun.

Not only are the tides interesting for their relation to navigation, but on account of their intimate connection with the motion of the Earth. The daily movement of the tides is a drag on the energy of our planet and is acting to cut down its speed of rotation. If the speed of rotation of the Earth diminishes it is obvious that our day is being lengthened, that to-day is longer than yesterday and that to-morrow will be longer than to-day, even tho these increases are all but inappreciable in the time that the human mind can fathom. Sir Robert Ball in his "Story of the Heavens" summarizes the variation wrought by the tides in the following paragraph: "Let us take a glance back into the profound depths of times past and see what the tides have to tell us. If the present order of things has lasted, the day must have been shorter and shorter the farther we look back into the dim past. The day is now twenty-four hours; it was once twenty hours, once ten hours, and once six hours. How much farther can we go? When the six hours it past, we begin to approach a limit which must at some point bound our retrospect. The shorter the day the more is the Earth bulged at the equator; the more the Earth is bulged at the equator the greater is the strain put upon the materials of the Earth by the centrifugal force of its rotation. If the Earth were to go too fast it would be unable to cohere together; it would separate into pieces, just as a grindstone driven too rapidly is rent asunder with violence. Here, therefore, we discern in the remote past a barrier which stops the present argument. There is a certain critical velocity which is the greatest that the Earth could bear without risk of rupture, but the exact amount of that velocity is a question not very easy to answer. It depends on the nature of the materials of the Earth; it depends upon the temperature; it depends upon the effect of pressure and on other details not accurately known to us. An estimate of the critical velocity has, however, been made, and it has been shown mathematically that the shortest period of rotation that the Earth could have, without flying into pieces, is about three or four hours. The doctrine of tidal evolution has thus conducted us to the conclusion that at some inconceivably remote epoch the Earth was spinning around its axis in a period approximating to three or four hours."

The existence of tides in the solid Earth, as well as in the oceans, was first indicated by Lord Kelvin in 1867. The first numerical estimate of the amount of the tidal oscillations of the solid Earth was made by Sir George H. Darwin in 1882 and was based on an indirect method and resulted from observation of the tides of the ocean. Further indirect methods were successfully applied later, with more complete data.

Sir George H. Darwin has expressed the opinion that we may now feel confident that the Earth yields to the tidal forces to the same degree as would a steel globe. The amazing accuracy of Dr. Hecker's recent observations (1908) are marked by certain collateral considerations, among which may be mentioned the flexure of coast lines, due to the varying pressure of oceanic tides, and the effects of high and low barometric pressure in bending the Earth's surface. It is impossible to give an accurate measurement of the amount of the vertical motion of the Earth's surface, but it is probable that at spring tides in this latitude the surface of the solid Earth moves up and down through about six inches.

CHAPTER XIV

THE MOON

That the Moon is the nearest the Earth of all the heavenly bodies is one of the most obvious facts which confront the star-gazer. Its regular motion must have been early appreciated by primitive man, after he had realized that the rising and setting of the Sun marks regularly recurrent intervals of time. As he was able to reflect upon his observations and properly to coördinate them, he doubtless noted the connection between the form of the Moon and its position in the sky with respect to the Sun. For by keeping count with pebbles or rude notches cut in a stick he would learn that the interval of time between recurring full moons was always the same, and that the series of changes he could observe followed in regular order. Thus when the Moon appeared after sunset near the place where the Sun had disappeared he saw a thin crescent, the hollow side of which was turned away from the Sun. A little later the Moon set.

The next night he observed the Moon further from the Sun, with a thicker crescent, and noted also that it set later, an effect that gradually increased until the semicircular disk, with the flat side turned away from the Sun, was seen in rather less than a week after the first appearance of the crescent. In another week the semi-circle enlarged to a complete disk and the Moon rose about sunset and set about sunrise, being in a direction nearly opposite to that of the Sun. From this time on the size again

Direction from which the Sun's rays are coming.

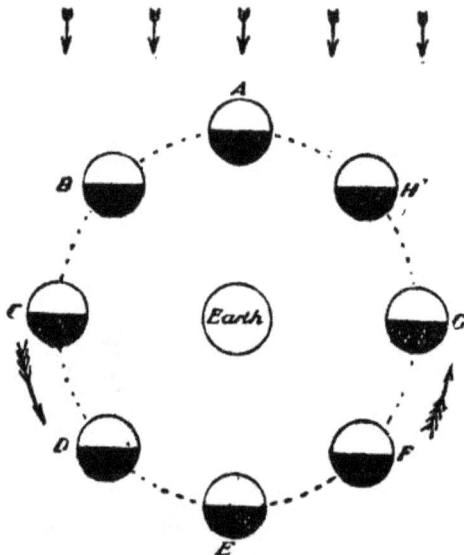

Various positions and illuminations of the Moon by the Sun during her revolution around the Earth.

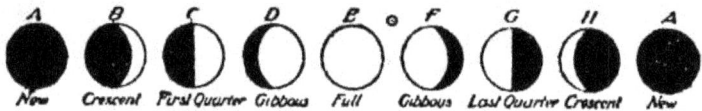

A	B	C	D	E	F	G	H	A
New	Crescent	First Quarter	Gibbous	Full	Gibbous	Last Quarter	Crescent	New

Fig. 21 —Orbit and Phases of the Moon.

The corresponding positions as viewed from the Earth, showing the consequent phases.

diminished; the semi-circular form was seen once more with the flat side still turned away from the Sun and toward the west instead of the east as the Moon approached the Sun on the other side, rising before it and setting in the daytime. Again he saw the crescent and marked that the time of rising approached that of sunrise, until the Moon became altogether invisible. Two or three nights intervened and the new Moon reappeared, whereupon the whole series of changes was repeated. In other words, this primitive man must have formulated a lunar month.

All ancient records recognise this lunar month of twenty-nine and a half days, and that interval must have been adopted long before the year. By the time of Chaldean and Egyptian astronomy, however, the year was known, so that the first conception of the lunar month is lost in the mists of antiquity. The Chaldeans studied the motions of the Sun and the Moon. Their calendar records, which they seem to have maintained with considerable care, enabled them to discover that eclipses occurred after a period known as the Saros, consisting of 6,585 days.

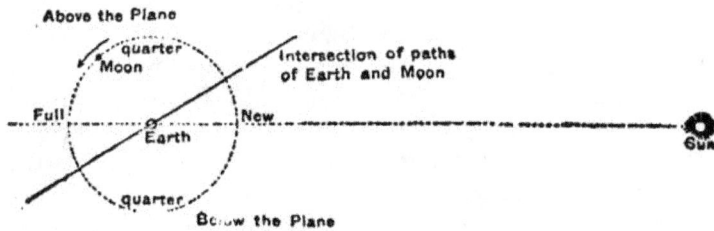

Fig. 22 —THE MOON'S PATH AROUND THE SUN.

The nature of the Moon was, of course, to them a mystery. It was known to move around the sky. The Babylonians supposed that, having a bright and a dark side, the different phases were caused by the bright side's coming more and more into view during its movement around the sky. In the seventh century Pythagoras taught more correctly that the Moon, like other heavenly bodies, was spherical, and that it was bright because it received the light of the Sun. The phases, he rightly judged, were due to a greater or less amount of the illuminated half turned toward us, and the curve forming the·boundary between the bright and dark portions of the Moon was to him conclusive evidence of a spherical shape. He was later supported in his theory by Aristotle, who made a similar clear and definite statement of the reason for the phases of the Moon.

Perhaps the first systematic study of the Moon was that of Aristarchus, a famous member of the Alexandrine school, who flourished in the first half of the third century and wrote a treatise "On the Magnitudes and Distances of the Sun and Moon," which still survives. Taking the Moon when it was half full, so that a line drawn to it from the Sun made a right angle with a line from the Moon to the Earth, by measuring the angle between the Moon and the Sun he was able to determine the ratio of the sides of this triangle and the relative distance of the Moon and the Sun from the Earth. While the method of Aristarchus was ingenious, yet the result he obtained, that the Sun was 18 to 20 times as far distant as the Moon, was sadly in error, for the actual distance is nearly four hundred times. Still the difficulties in the way of accurate observation were enormous for lack of proper instruments. Aristarchus also took advantage of the solar eclipse to ascertain the distances of these two bodies, and reasoned correctly that when the Moon sometimes rather more than hides the surface of the Sun and sometimes does not quite cover it, their diameters must vary as their real distances. He even obtained by eclipse observations a value for the diameter of the Moon in terms of that of the Earth. What would have been an excellent approximation was marred, however, by an incorrect estimate of the apparent size of the Moon, which he took as two degrees instead of one-half a degree. Nevertheless this work was an important advance in practical astronomy. It paved the way for Hipparchus, who discussed the motions of the Moon, motions which, long before his time, were known to be irregular and much more complicated than those of the Sun. The path of the Moon is always changing and its motions are subject to variation.

Hipparchus notes that the part of the Moon's path in which the motion is most rapid is not always in the same position on the celestial sphere but moves continuously. He was able to realize the different motions of the Moon, to distinguish the various months based upon them and to employ the Chaldean and other early eclipse observations for determining the position of the Moon at various earlier epochs. Hipparchus really evolved a most complicated set of motions. He was aware of the shortcomings of his theory, but was unable to reconstruct it in a satisfactory manner. Following out the eclipse method of Aristarchus, Hipparchus made a determination of the relation between the distances of the Sun and Moon, measuring their angular diameters, and found that the distance of the Moon is nearly 59 times the radius of the Earth. Combining this figure for the distance of the Sun with that of Aristarchus, a value of 1,200 times the radius of the Earth was obtained, which value was employed for many centuries.

Ptolemy, in the fourth book of the "Almagest," discusses the length of the month with the theory of the Moon and makes the important announcement of a further inequality in the Moon's motion, which Hipparchus had only suspected and which was due in large part to its position with respect to the Sun. This was later termed "evection" and involved a still further complication of the mathematical computations of Hipparchus because it meant the use of an epicycle and a deferent, which was itself a moving eccentric circle, the center of which revolved around the Earth. Ptolemy's mathematics and ingenuity were able to fit his theory to observations. Altho his work showed many inconsistencies which, great as he was, he was unable to control, nevertheless it represented a notable development in astronomy. To Ptolemy is due the parallax method for obtaining the distance of the Moon by observing its direction from two points on the Earth's surface and by finding the distance between these two points in terms of the Earth's radius. This distance of the Moon he estimated was 59 times the radius of the Earth. With this value, according to the method of Hipparchus just mentioned, he computed the distance from the Earth to the Sun. He found that the distance was 1,210 times the radius of the Earth, which value was equally in error as compared with the modern figure. The actual distance is about twenty times this amount.

So readily were the motions of the Moon observed and so carefully were the records maintained that even at an early date much was known about its motions. These data enabled much calculation of the Moon's motion to be carried on, so that early mathematical astronomy dealt with the Moon in no small degree. Various irregularities of its motion and appearance were discussed, and practically all of the notable astronomers made some contributions to our knowledge of the satellite. Even an outline of these discussions would lead the reader far afield into the depths of mathematics, for which reason it is possible to mention only a few of the important researches and to indicate merely their general nature. The study of the Moon had a practical importance, however, as lunar positions were early used to determine longitude in navigation and lunar tables were calculated by government and other astronomers.

Nowhere was there more interest manifested in the problem of the motion of the Moon than at Greenwich Observatory, where the matter had always been a specialty of that institution. "It is a curious fact," states the late Professor Newcomb in his "Reminiscences of an Astronomer," "that while that observatory supplied all the observations of the Moon, the investigations based upon these observations were made almost entirely by foreigners, who also constructed the tables by which the Moon's motion was mapped out in advance. The most perfect tables made were those of Hansen, the greatest master of mathematical astronomy during the middle of the century, whose tables of the Moon were published by the British Government in 1857. They were based on a few of the Greenwich observations from 1750 to 1850. The period began with 1750, because that was the earliest at which observations of any exactness were made. Only a few observations were used, because Hansen, with the limited computing force at his command—only a single assistant, I believe—was not able to utilize a great number of the observations. The rapid motion of the Moon, a circuit being completed in less than a month, made numerous observations necessary, while the very large deviations in the motion produced by the attraction of the Sun made the problem of the mathematical theory of that motion the most complicated in astronomy. Thus it happened that, when I commenced work at the Naval Observatory in 1861, the question whether the Moon exactly followed the course laid out for her by Hansen's tables was becoming of great importance.

"For a year or two our observations showed that the Moon seemed to be falling a little behind her predicted motion. But this soon ceased, and she gradually forged ahead in a much more remarkable way. In five or six years it was evident that this was becoming permanent; she was a little farther ahead every year. What could it mean? To consider this question, I may add a word to what I have already said on the subject.

"In comparing the observed and predicted motion of the Moon, mathematicians and astronomers, beginning with Laplace, have been perplexed by what are called 'inequalities of long period.' For a number of years, perhaps half a century, the Moon would seem to be running ahead and then she would gradually relax her speed and fall behind. Laplace suggested possible causes, but could not prove them. Hansen, it was supposed, had straightened out the tangle by showing that the action of Venus produced a swinging of this sort in the Moon; for one hundred and thirty years she would be running ahead and then for one hundred and thirty years more falling back again, like a pendulum. Two motions of this sort were combined together. They were claimed to explain the whole difficulty. The Moon, having followed Hansen's theory for one hundred years, would not be likely to deviate from it. Now it was deviating. What could it mean?

"Taking it for granted, on Hansen's authority, that his tables represented the motions of the Moon perfectly since 1750, was there no possibility of learning anything from observations before that date? As I have already said, the published observations with the usual instruments were not of that refined character which would decide a question like this." But observations of stars which might be available, and which had been made in many instances, would have an important bearing on this work. For if an occultation or passage of the Moon between a star and the Earth occurred and its time registered, the path of the Moon in the heavens and the time at which it arrives at each point of the path would be determined if the time of the occultation was known within one or two seconds. Professor Newcomb continues: "It was not until after the middle of the century (seventeenth), when the telescope had been made part of astronomical instruments for finding the altitude of a heavenly body, and after the pendulum clock had been invented by Huygens, that the time of an occultation could be fixed with the required exactness. Thus it happens that from 1640 to 1670 somewhat coarse observations of the kind are available, and after the latter epoch those made by the French astronomers become almost equal to the modern ones in precision.

"The question that occurred to me was, Is it not possible that such observations were made by astronomers long before 1750? Searching the published memoirs of the French Academy of Sciences and the Philosophical Transactions, I found that a few such observations were actually made between 1660 and 1700. I computed and reduced a few of them, finding with surprise that Hansen's tables were evidently much in error at that time. But neither the cause, amount or nature of the error could be well determined without more observations than these. Was it not possible that these astronomers had made more than they had published? The hope that material of this sort existed was encouraged by the discovery at the Pulkowa Observatory of an old manuscript by the French astronomer Delisle, containing some observations of this kind. I therefore planned a thoro search of the old records in Europe to see what could be learned."

By good fortune suitable observations were found at the Paris Observatory and their relations and the method of making were studied and everything necessary was copied. This work took some six weeks, but Professor Newcomb says, "The material I carried away proved the greatest find I ever made. Three or four years were spent in making all the calculations I have described. Then it was found that seventy-five years were added, at a single step, to the period during which the history of the Moon's motion could be written. Previously this history was supposed to commence with the observations of Bradley at Greenwich, about 1750; now it was extended back to 1675, and with a less degree of accuracy thirty years farther still. Hansen's tables were found to deviate from the truth in 1675 and subsequent years to a surprising extent, but the cause of the deviation is not entirely unfolded even now.

"One curious result of this work is that the longitude of the Moon may now be said to be known with greater accuracy through the last quarter of the seventeenth century than during the ninety years from 1750 to 1840. The reason is that, for this more modern period, no effective comparison has been made between observations and Hansen's tables."

As the Moon rotates around the Earth while the Earth is·passing in its orbit about the Sun, it will be obvious that twice in its journey the Sun must come into a line of intersection of the Moon's orbit with that of the Earth.

When this happens at the time of full moon the Earth will lie directly between the Moon and the Sun, so that the light from the Sun is intercepted and a shadow is formed on the Moon's surface. Sometimes the Moon only partially enters the Earth's shadow and then the eclipse, as this phenomenon is termed, is partial. If, however, the Sun is situated on the line of intersection at the time of new moon, then the Sun will be eclipsed, and a solar eclipse so rich in astronomical significance occurs. Solar eclipses have been discussed elsewhere and their importance explained. Lunar eclipses, besides enabling us

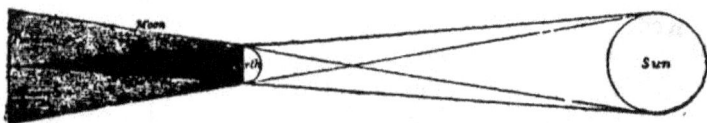

Fig. 23 —FORM OF THE EARTH'S SHADOW, SHOWING THE PENUMBRA, OR PARTIALLY SHADED REGION. WITHIN THE PENUMBRA THE MOON IS VISIBLE; IN THE SHADOW IT IS NEARLY INVISIBLE.

to check the motions of the Moon and furnishing an interesting spectacle, afford little scientific information.

When the black shadow on the Moon is first detected in the case of a total lunar eclipse, it is interesting to watch its encroachment until the entire surface of the satellite is covered. Even the Moon, on which no direct sunlight can fall, is often visible, glowing with a copper-colored hue, sufficiently bright to enable several of the markings on the surface to be seen. This is due to refraction of the atmosphere, which bends the sunbeams that have just grazed the Earth and permits them to fall within the shadow. In their journey through the denser atmosphere they have become rich in red rays, which gives to the disk a ruddy or copper-like hue analogous to that of the Sun at sunrise or sunset. Whether the eclipse of the Moon is total or partial depends on the extent to which it passes into the shadow of the Earth, as the accompanying diagram will indicate clearly. Lunar eclipses are useful to the astronomer for determining the length of the synodic month and also for determining the temperature to which the Moon has been raised, for when it enters the shadow all direct light from the Sun is cut off and the Moon becomes cold very rapidly. Furthermore, the position of the Moon with respect to the stars can be determined on such occasions with great accuracy. Like solar eclipses, eclipses of the Moon can be predicted with high precision, and they are regularly announced in almanacs and ephemerides.

The Moon always presents the same face to the Earth, a phenomenon discovered by Galileo. It must follow that the Moon rotates on its axis once in the same number of seconds that it requires for a revolution around our planet. This is explained by the fact that tides on the Moon, as in the case of the Earth, have lengthened the period of rotation by their braking action. At a time when the Moon was still a hot, semi-molten mass, the attraction of the Earth produced great tides, not tides of water, but tides of molten rock These tides on the Moon checked its rotational velocity and eventually constrained the Moon to rotate on its axis in precisely the same period as that which it requires to revolve around the Earth. All this happened eons ago. There is no longer evidence of any tidal action, because the Moon is frozen. Altho there can hardly be tides on the Moon, yet there may be tides in the Moon.

"It may be that the interior of the Moon is still hot enough to retain an appreciable degree of fluidity," writes Sir Robert Ball, "and if so, the tidal control would still retain the Moon in its grip; but the time will probably come, if it have not come already, when the Moon will be cold to the center—cold as the temperature of space. If the materials of the Moon were what a mathematician would call absolutely rigid, there can be no doubt that the tides could no longer exist, and the Moon would be eman-cipated from tidal control. It seems impossible to predi-cate how far the Moon can ever conform to the circum-stances of a rigid body, but it may be conceivable that at some future time the tidal control shall have practically ceased. There would then be no longer any necessary identity between the period of rotation and that of revolu-tion. A gleam of hope is thus projected over the astron-omy of the distant future. We know that the time of revolution of the Moon is increasing, and so long as the tidal governor could act, the time of rotation must increase sympathetically. There will then be nothing to prevent the rotation remaining as at present, while the period of revolution is increasing. The privilege of seeing the other side of the Moon, which has been withheld from all previous astronomers, may thus in the distant future be granted to their successors." .

While study of the Moon and its motions continued, a beginning was made in the Renaissance to examine the surface of this satellite. Leonardo da Vinci (1452-1519) was the first to explain correctly the dim illumination seen over the rest of the surface of the Moon when the bright part was only a thin crescent. This he maintained is due to the earthshine or slight illumination of the Moon by light reflected from the Earth, just as moonshine is able to illuminate the Earth.

Galileo's lunar observations through his telescope were epoch-making. Not only was he able to disprove many common conceptions of the nature of the satellite and its surface, but also to present a mass of evidence of a positive character. In spite of the familiar dark markings, the Moon was really supposed to be a smooth sphere. After the introduction of the telescope, however, it was recog-nised by Galileo that the surface of the Earth's satellite was dotted with various inequalities, which he assumed to be mountains, valleys and seas. Thus he correctly ac-counted, in part at least, for the unevenness of the surface.

He was not content with mere observation of the features
of the Moon's surface, but measured the height of some
of the more conspicuous lunar mountains and obtained
for them an estimated elevation of four miles, a figure
which agrees fairly well with modern estimates. Having
seen and measured mountains on the Moon's surface, it
seemed natural that there should be water. The large
dark spots he erroneously regarded as seas, altho he was
not responsible for the corresponding names applied to
these supposed expanses of water by some of his successors, and still preserved in lunar maps. The chief
marks of astronomical progress as revealed by Galileo's
observation of the Moon were that it was a body in many
respects similar to the Earth, that it was not a perfect
sphere, and that there is no fundamental difference between celestial bodies and our own Earth, either in their
motions or in their general nature, which was important
in the final establishment of the Copernican theory.

One other discovery of Galileo's in connection with the
Moon is of great importance. It had been known for
many years that as the Moon revolved around the Earth
the same markings were constantly seen. With the telescope these markings could be studied so much more distinctly that it occurred to Galileo to ascertain whether
there was any change in the Moon's disk, or whether its
appearance was always exactly the same. He found that
as the Moon moves in its orbit around the Earth that
slight changes are seen in its appearance. In other words,
small portions of the hemisphere alternately on its northern and southern half are exposed. The simplest of the
motions of the Moon in this way subsequently came to be
known as "librations."

Kepler in his Epitome of the Copernican astronomy
demonstrated that his planetary laws applied to the motion
of the Moon around the Earth, despite irregularities
which introduced enormous complications. In this work,
however, he devotes much attention to the theory of the

Moon, explaining in considerable detail both evection and variation.

Galileo established the fact that the Moon was similar to the Earth in many respects. The analogy was carried somewhat further by certain of the pioneer workers with large telescopes. Even Herschel held that because the Moon closely resembled the Earth, it might be a suitable habitat for human beings. The dark spots once taken for seas and bearing that name on lunar maps are, in reality, lava, while the craters which dot the surface of the satellite, with one or two possible exceptions, belong to volcanoes long since extinct. The dark lines once known as "rills," which it was assumed were rivers, are plainly without water. If there is a lunar atmosphere its density must be very small, in fact less than that of the atmosphere far above the Earth. That there is a very rare lunar atmosphere seems to be probable. In fact, the assumption of an atmosphere is necessary for the explanation of certain phenomena.

After Galileo's lunar studies the next important work was that of John Hevel, of Danzig (1611-1687), who published in 1647 his Selenographia, in which not only the text but the plates were prepared by him. Here he systematically describes and names the chief features of the Moon, the immense craters and seas, employing many names taken from the Earth, such as Apennines, Alps, and Mare Serenitatis for the Pacific Ocean, all designations still found on modern lunar maps. Not all of his names have survived. John Baptist Riccioli (1598-1671), in a treatise on astronomy called The New Almagest, 1651, gave to various mountains and craters the names of distinguished men of science and philosophers. Hence the names of Plato, Archimedes, Tycho and Copernicus are found on lunar maps. These names have survived in considerable number. More modern map-makers, such as Beer and Maedler, whose map was published in 1837, and Schmidt of Athens, who published a map of the Moon seven feet in diameter in 1878, have carried out this idea. Modern astronomers are likewise honored with the names of various points represented on the maps.

For the origin of the Moon the mind must be forced to look back millions and millions of years to a time when our Earth existed in a very different form. Then it was not a solid mass, but a globe of molten material on which floated a crust perhaps some thirty-five miles thick. It rotated not in a period of 24 hours, the present day's length, but at a terrific velocity which may have made the day some three hours in length. Such a speed of revolution naturally produced a most powerful centrifugal force. One day a cataclysm occurred. Five thousand cubic millions of miles of matter were thrown off into space. Thus the Moon was born. A great scar was left on the surface of the Earth, a scar which in the opinion of Professor William H Pickering is the basin of the Pacific Ocean. After this rending of the Earth the remaining parts of the crust, afloat on the liquid interior, were split along irregular lines into two pieces, which drifted apart and were filled by the waters of the Atlantic and Pacific Oceans. This theory of Pickering's is diametrically opposed to that of Professor T. J. J. See, who claims that the Moon is in reality a planet, captured by the Earth in its wanderings through space, and that all satellites have been thus captured.

Whatever the origin of the Moon, it is the largest of all the planetary satellites, yet smaller in mass than the Earth, from which it is separated by a distance that varies between 222,000 and 253,000 miles. Its gravity is equal to about one-sixth that of the Earth, for which reason the same amount of energy acting, for example, in a volcanic upheaval, produced mountains higher than those on the Earth. This mass of the Moon is about $\frac{1}{80}$ that of the Earth, or 73,000,000,000,000 tons.

The accompanying diagram shows the comparative size of the Earth and the Moon. The diameter of our planet is 7,914 miles, while that of the Moon is 2,160 miles, so that the diameters stand very nearly in the relation of 4 to 1, while the superficial area of the Moon is equal to about $^1/_{13}$ part of the surface of the Earth. The average distance of the Moon from the Earth is also fairly constant, and the average fluctuations do not exceed more than about 13,000 miles on either side of its mean value of 239,000 miles.

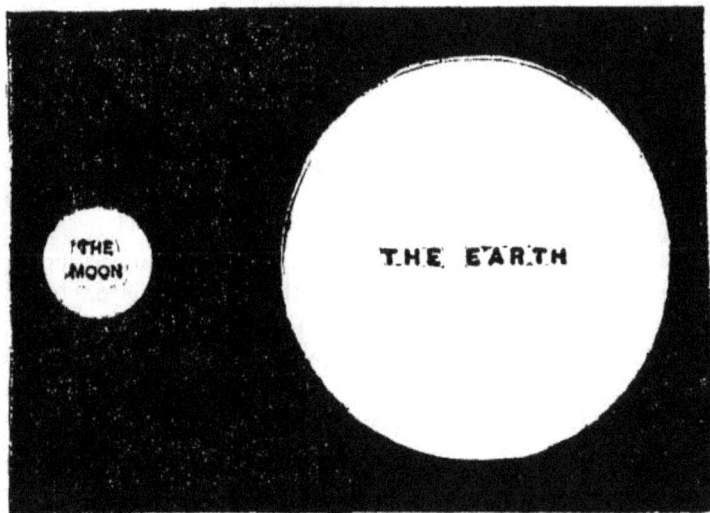

Fig. 24 —Comparative Sizes of the Earth and the Moon.

The Moon is essentially a dead planet in the eyes of most astronomers. Its fires long since have been extinguished. It is a great globe of chilled slag. Its craters have no counterpart on the Earth. The lunar crater is a great circular plane, 50 or even 100 miles in diameter, around which rises a precipice perhaps five or ten thousand feet, while in the center there may be a hill or two about half as high. Thousands of these volcanoes are visible in the telescope. How these craters were formed is a puzzle. Some astronomers hold that they mark the impact of countless meteorites. Others assume them to be the product of gigantic bubbles in a once molten mass that burst. Again it is claimed that they are volcanoes resembling those of the Earth

Most astronomers tell us that the craters have long been dead, that the Moon has had for centuries no atmosphere and therefore cannot have water or support plant or animal life. Prof. William H. Pickering, however, tells us of an exceedingly rare lunar atmosphere and maintains that the Moon's craters are not all extinct. He even claims that certain great white expanses are snow and ice, and, furthermore, that there is evidence of the growth and decay of vegetation. To support his views of vanishingly feeble volcanic activity he calls attention to the little crater named Linné, after the famous naturalist, Linnæus. In modern times this crater has unquestionably undergone changes. Only a few centuries ago it was described on the old maps as "a crater of moderate size" and later as "a very small round brilliant spot." A dead volcano cannot alter its shape.

If there are still a few intermittently active volcanoes they must expel water and carbonic acid gas, judging by earthly volcanoes. But water cannot be found on the Moon as a liquid, for the temperature of its surface is probably not far from 460° F. below zero. Many of the elevated peaks are capped with a silver glow, which also characterizes the lining of the "craterlets." In Pickering's eyes that white sheen is snow. He believes also in accumulations of snow and ice at the poles. If carbonic acid is expelled by lunar volcanoes it must cling to the Moon with great tenacity because of its weight. Since it is the food of plant life, it may possibly support vegetation on the Moon. Professor Pickering sees in variable dark spots on the planet organic life resembling vegetation. He figures that since certain lichens grow in certain regions of the Earth where the temperature never rises above the freezing point, there is no reason why that vegetation may not flourish on the Moon's surface.

CHAPTER XV

MARS

MARS is another instance of a celestial body whose discovery long antedates written history, for which reason it is natural to assume that its appearance in the heavens and its motions have always been known to mankind. Its ruddy light naturally associated it in the Greek mind with the God of War, just as the soft, warm glow of Venus was considered appropriate to the Goddess of Love. To the Chaldean and to other ancient star-gazers the motions of the planet were indeed of interest. No planet has received greater attention from astronomers or has been used more for the solution of important astronomical and cosmical problems.

Thus it was from an observation of the passage of the Moon in front of Mars, or what is termed by astronomers an occultation of the planet, that Aristotle concluded that the planets are more distant than the Sun and Moon. What reason he had for including the Sun it is difficult to conceive, as an occultation by that body would be quite impossible. Later it was from the study of the motions of the planet made for Tycho Brahe that Kepler was able to derive his great laws of planetary motion. By observing the great "Hour-glass Sea," or Syrtis Major, on Mars, Huygens in 1659 noted the rotation of the planet on its axis, and from a measurement of the planet made at its opposition Richer was able to obtain a reasonably accurate estimate of the distance of the Sun in 1673.

In 1783 Herschel had written: "The analogy between Mars and the Earth is perhaps by far the greatest in the whole solar system." For over a century the development of this analogy has been one of the most important features of astronomy, especially in popular estimation. Today, with a wealth of scientific and observational data acquired during the last score of years, the question, altho more fully and intelligently discussed, is as far from solution as ever.

Mars revolves around the Sun in an orbit whose mean distance is 141,500,000 miles. But this orbit is so eccentric, amounting to c.093, that the distance varies about thirteen million miles. As this distance from the Sun is about one-half as far again as that of the Earth, the light and heat that Mars receives are only about $^4/_9$ of the light and heat received by the Earth, an important consideration in discussing the possibility of life on its surface.

The real diameter of Mars is about 4,200 miles, a figure correct within 20 miles, so that its surface is $^{28}/_{100}$ and its volume 0.147 or $^1/_7$ that of the Earth. The mass of the planet, derived from observations of its satellites, is about $^1/_{9.4}$ that of the Earth, so that its density is 0.73 and the attraction of gravity on its surface 0.38 that of the Earth. Mars reflects 0.26 of the light that it receives, which is just double that of Mercury.

In Mars may be seen a striking difference between the superior and inferior planets, for the crescent phases of the latter are conspicuously absent. This is because the orbits of Mars and the other superior planets lie outside of the Earth. Consequently such a planet never comes between the Earth and the Sun. At quadrature a certain amount of the unilluminated portion is turned toward the Earth so that the planet appears like a Moon three or four days from full, with a distinctly gibbous appearance.

·Kepler suspected the existence of satellites from a somewhat cryptic utterance of Galileo's in regard to the triple Saturn. No particularly sound reasons were advanced for the existence of such satellites. Still that they might be found is forcibly reflected in eighteenth-century literature. In "Gulliver's Travels" (1726) is the striking statement that an astronomer on the floating island of Laputa had discovered two tiny satellites of Mars. Dean Swift even advanced the extraordinary statement that one of these moons revolved around the planet in 10 hours, the approximate correctness of which will presently be apparent. Furthermore, Voltaire in his "Micromégas," published in 1752 and obviously modeled on "Gulliver's Travels," also speaks of these small bodies. Altho literature and imagination had preceded observation, astronomers could not detect any companions to the ruddy planet until August 11, 1877, when Professor Asaph Hall, using the 26-inch refractor of the U. S. Naval Observatory at Washington, which at the time of its construction (1873) was the largest telescope in the world, studied the vicinity of Mars with unusual care. In 1877 it was at conjunction, so that conditions were favorable for an exhaustive inquiry. On August 11 Professor Hall detected a minute speck of light near the planet, and a few days later, August 16, he ascertained positively that it was a satellite. On the following evening a second body was discovered, bewilderingly rapid in its motion. Since that time the two satellites have been frequently observed at opposition even with much smaller instruments. On July 18, 1888, the large Lick telescope was able to reveal them when their brightness was only about one-eighth that at the time of their discovery.

The names appropriately assigned to the new satellites, Deimos and Phobos (Fear and Panic), were taken from the Iliad and represent the companions in battle of the God of War. The satellites are very small, having diameters respectively of six and seven miles, as photometrically measured at Harvard University soon after their discovery by Professor Pickering. Phobos, the inner, is the larger and traverses an elliptical orbit in 7 h. 39 m. 22 s. at a distance of only 3,760 miles. Deimos, its companion, moves in a nearly circular orbit in 30 h. 18 m. at a distance from the planet of 12.500 miles.

It should be remembered in all discussions of the nature of the surface of the planet Mars and in developing theories based thereon that astronomers must deal with a little disk which once in fifteen years has a maximum of about $1/_{5000}$ the area of the full Moon. When nearest the Earth Mars is at a distance of 35,500,000 miles and when observed with a telescope magnifying 1,000 times it appears only as large as the Moon with an ordinary field glass, enlarging six or seven times. In other words, Mars is brought to a distance of 35,500 miles. Hence it is difficult to distinguish minute details. Consequently visual and photographic observation of Mars must be of the most refined character, for especially in the former there is opportunity for psychological or subjective phenomena to occur in connection with the observation. While it is neither rational nor scientific to deny what credible observers claim they have seen through the telescope in examining the planet, yet it must be recognised that visual observations are attended by necessary limitations and should be received in a spirit of caution rather than of enthusiasm.

The mapping of Mars is no recent matter, for even in 1659 a rough sketch of the surface of the planet was made by Huygens in which the V-shaped marking at the equator pointing to the north can be identified as the Syrtis Major. This was followed by rough sketches from time to time down to 1840, when Maedler first began a systematic charting of the planet. His map was followed in 1864 by Kaiser's, by Flammarion's in 1876 and Green's in 1877. Drawings of various parts of the planet were made during these intervals, but were not combined into good charts.

Observations made by Professor Lowell and his staff at the observatory at Flagstaff, Arizona, have had the study of this planet especially in view. The Lowell observatory itself is situated 7,300 feet above the sea and contains an excellent 24-inch telescope with which it is claimed that the faintest stars shown on the chart made at the Lick Observatory with a 36-inch instrument are perfectly visible. The atmosphere is well adapted for telescopic work, so that Professor Lowell and his colleagues possess singular advantages over other observers. On the other hand, in many important respects there has been a lack of corroboration of their observations.

The surface of Mars, as seen in the telescope, is composed of two white polar caps, which wane with the approach of summer; orange areas, which are supposed by Lowell and his followers to be deserts, and blue-green areas, which change their hue to orange during the Martian autumn and winter and reassume their verdant tint in spring. The planet is covered with a network of fine lines, first discovered by Schiaparelli in 1877 and called by him "canals," a designation by which they are still known. These canals connect the polar caps with the temperate and equatorial zones. According to Professor Lowell, they may be regarded as planetary irrigation ditches which serve the purpose of leading the melting water of the poles to those desert regions which would still blossom if properly watered. The canals disappear with the approach of winter and creep down from the poles toward the equator in summer, a phenomenon which long puzzled astronomers until Pickering ingeniously suggested that we see not the canals themselves (for they are much too narrow), but the vegetation which fringes their banks, which withers as the cold of winter descends and which flourishes with the melting of the snows.

It was Schiaparelli who also first announced the curious doubling of the canals in certain instances, an announcement which was at first received with derision but which has since been confirmed by his disciple, Lowell, and the astronomers of the great observatories. This curious gemmation has not been satisfactorily explained, altho Professor Lowell attributes it in part to vegetal causes. By opponents of his theory the doubling is regarded as an optical illusion, which can hardly be the case, because all and not certain of the canals would exhibit the phenomenon. Furthermore, there is some photographic evidence offered by Mr. Lampland, of Professor Lowell's staff, that both the canals and their doubling are real.

The visual observations of Professor Lowell have been undertaken with great thoroness and have brought to light much material which he has employed in the development of his theories. To do justice to his work in a single chapter is manifestly impossible. It will be most satisfactory to the reader to recapitulate in Professor Lowell's own words, derived from his work on "Mars and Its Canals," the fundamental facts which he has recorded and on which he bases his conclusions. They are as follows:

"(1) Mars turns on its axis in 24 h. 37 m. 22.65 s. with reference to the stars and in 24 h. 39 m. 35 s. (as a mean) with regard to the Sun. Its day therefore is only about forty minutes longer than ours.

"(2) Its axis is tilted to the plane of its orbit by about 23° 59′ (most recent determination, 1905). This gives the planet seasons almost the counterpart of our own in character, but in length nearly double ours, for

"(3) Its year consists of 687 of our days, 669 of its own.

"(4) Polar caps are plainly visible which melt in the Martian summer to form again in the Martian winter, thus implying the presence of a substance deposited by cold.

"(5) As the polar caps melt they are bordered by a blue belt, which retreats with them. This excludes the possibility of their being formed of carbon dioxide and shows that of all the substances we know the material composing them must be water.

"(6) In the case of the southern cap, the blue belt has widenings in it in places. These occur where the blue-green areas bordering upon the polar cap are largest.

"(7) The extensive shrinkage of the polar snows shows their quality to be inconsiderable and points to scanty deposition due to dearth of water.

"(8) The melting takes place locally after the same general order and method, Martian year after year, both in the south cap

"(9) And in the north one. This is evidenced by the recurrence of rifts in the same places annually in each. The water thus let loose can, therefore, be locally counted on.

"(10) That the south polar cap is given to greater extremes than the north one implies again, in view of the eccentricity of the orbit and the tilt of the axis, that deposition in both caps is light.

"(11) The polar seas at the edges of the caps being temporary affairs, the water from them must be fresh.

"(12) The melting of the caps on the one hand and their re-forming on the other affirm the presence of water vapor in the Martian atmosphere, of whatever else that air consist.

"(13) Since water vapor is present, of which the molecular weight is 18, it follows from the kinetic theory of gases that nitrogen, oxygen and carbonic acid, of molecular weights 28, 32 and 38 respectively, are probably there, too, owing to being heavier.

"(14) The limb-light bears testimony to this atmosphere.

"(15) The planet's low albedo points to a density for the atmosphere very much less than our own.

"(16) The apparent evidence of a twilight goes to affirm this.

"(17) Permanent markings show upon the disk, proving that the surface itself is visible.

"(18) Outside of the polar cap the surface is divided into red-ocher and blue-green regions. The red-ocher stretches have the same appearance as our deserts seen from afar,

"(19) And behave as such, being but little affected by change.

"(20) The blue-green areas were once thought to be seas. But they cannot be such because they change in tint, according to the Martian season, and the area and the amount of the lightening is not offset at the time by corresponding darkening elsewhere

"(21) Nor by any augmentation of the other polar cap or precipitation into cloud. It cannot, therefore, be due to shift of substance.

"(22) Furthermore, they are all seamed by lines and spots darker than themselves, which are permanent in place, so that there can be no bodies of water on the planet.

"(23) On the other hand, their color, blue-green, is that of vegetation. This regularly fades out, as vegetation would, to ocher for the most part, but in places changes to a chocolate-brown.

"(24) The change that comes over them is seasonal in period, as that of vegetation would be.

"(25) Each hemisphere undergoes this metamorphosis in turn.

"(26) That it is recurrent is again proof positive of an atmosphere.

"(27) The changes are metabolic, since those in one direction are later reversed to a restoration of the original status. Anabolic as well as katabolic processes thus go on there—that is, growth as well as decay takes place. This proves them of vegetal origin.

"(28) The existence of vegetation shows that carbonic acid, oxygen and undoubtedly nitrogen are present in the Martian atmosphere, since plants give out oxygen and take in carbonic acid.

"(29) The changes in the dark areas follow upon the melting of the polar caps, not occurring until after that melting is under way,

"(30) And not immediately then, but only after the lapse of a certain time.

"(31) Tho not seas now, from their look the dark areas suggest old sea bottoms, and when on the terminator appear as depressions (whether because really at a lower level or because of less illumination is not certain).

"(32) That they are now the parts of the planet to support vegetation hints the same past office, as water would naturally drain into them. That such a metamorphosis should occur with planetary aging is in keeping with the kinetic theory of gases.

"(33) Terminator observations prove conclusively that there are no mountains on Mars, but that the surface is surprisingly flat.

"(34) But they do reveal clouds which are usually rare and are often, if not always, dust-storms.

"(35) White spots are occasionally visible, lasting unchanged for weeks in the tropic and temperate regions, showing that the climate is apparently cold,

"(36) But at the same time proving that most of the surface has a temperature above the freezing-point.

"(37) In winter the temperate zones are more or less covered by a whitish veil, which may be hoar-frost or may be cloud.

"(38) A spring haze surrounds the north polar cap during the weeks that follow its most extensive melting.

"(39) Otherwise the Martian sky is perfectly clear, like that of a dry and desert land."

These facts, according to Professor Lowell, make reasonably evident on Mars the presence of—

"(1) Days and seasons substantially like our own.

"(2) An atmosphere containing water vapor, carbonic acid and oxygen.

"(3) Water in great scarcity.

"(4) A temperature colder than ours, but above the Fahrenheit freezing-point, except in winter and in the extreme polar regions.

"(5) Vegetation."

While all of Professor Lowell's observations and results

are not accepted by all astronomers, and there is considerable opposition to his conclusions, nevertheless they are of interest and worth stating as representing this aspect of the matter in his own words. Comparative studies of lunar and Martian spectra made on the summit of Mt. Whitney in September, 1909, by Campbell of the Lick Observatory seem to preclude the possibility of much water on Mars. Campbell's photographs show that the Martian atmosphere is no richer in water than the Moon's, which if true summarily disposes of Martian life. Slipher of Lowell's staff claims to have obtained ample spectroscopic evidence of water. The following paragraphs, taken from the concluding chapter of Lowell's book on "Mars and Its Canals," may be said fairly to sum up his views on Martian life:

"That Mars is inhabited by beings of some sort or other we may consider as certain as it is uncertain what those beings may be. The theory of the existence of intelligent life on Mars may be likened to the atomic theory in chemistry, in that in both we are led to the belief in units which we are alike unable to define. Both theories explain the facts in their respective fields and are the only theories that do, while as to what an atom may resemble we know as little as what a Martian may be like. But the behavior of chemic compounds points to the existence of atoms too small for us to see, and in the same way the aspect and behavior of the Martian markings implies the action of agents too far away to be made out.

"One of the things that makes Mars of such transcendent interest to man is the foresight it affords of the course earthly revolution is to pursue. On our own world we are able to study only our present and our past; in Mars we are able to glimpse, in some sort, our future. Different as the course of life on the two planets undoubtedly has been, the one helps, however imperfectly, to better understanding of the other."

The views expressed by Professor Lowell in the work

just quoted were further developed by him in the course of a few years succeeding its publication, and in "Mars as the Abode of Life," published in 1908, he expresses himself as even more firmly convinced that Mars is inhabited by a race of intelligent beings. Additional study of Martian phenomena, according to Professor Lowell, indicates that the canals and oases as he sees them are proof that life of no mean order of intelligence prevails on the planet. He suggests:

"Part and parcel of this information is the order of intelligence involved in the beings thus disclosed. Peculiarly impressive is the thought that life on another world should thus have made its presence known by its exercise of mind. That intelligence should thus mutely communicate its existence to us across the far stretches of space, itself remaining hid, appeals to all that is highest and most far-reaching in man himself. More satisfactory than strange this, for in no other way could the habitation of the planet have been revealed. It simply shows again the supremacy of mind. Men live after they are dead by what they have written while they are alive, and the inhabitants of a planet tell of themselves across space as do individuals athwart time by the same imprinting of their mind."

To this he adds the statement that conditions for life on the planet are approaching an end, as it is rapidly drying up, and that the energies of the inhabitants are being the slowly diminishing supply of water. "The drying up of the planet is certain to proceed until its surface can support no life at all." So that, as compared with the Earth, Mars presents a distinctly later period of evolution. No such danger at present confronts our own planet. But assuming that these explanations are correct, it is not improbable that the coöperative action of all the nations of the world may be required at some future date to deal with a similar problem.

In opposing the idea of canals on the surface of Mars, much stress has been laid on the alleged fact that the

finer markings and some of the apparent doublings are
based upon optical illusions or psychological phenomena.
Thus, to prove that instinctively the eye would arrange
in the form of straight lines vague suggestions of markings, E. W. Maunder, of the Greenwich Observatory,
England, and J. E. Evans, of the Royal Hospital School
at Greenwich, in 1902 performed an experiment with a
number of schoolboys. A circular disk was exhibited
five or six inches in diameter, on which was represented
with some accuracy the shaded area of the planet as it
might appear in the telescope. Instead of canals, a few
faint, wavy lines and a larger number of faint dots were
inserted promiscuously. The boys were told to fill in with
pencil on small circles on sheets of paper the details of the
object exhibited to them, exercising as much care as possible. The object of the experiment was not communicated
to the boys. None of them had any idea of the nature of
the planet's surface. The result was that boys at the
greatest distance from the object, where the faint lines
and dots were just beyond the limits of separate visibility,
drew canals strikingly like those noted by the telescopic
observers of the planet. Hence it was supposed that
the eye in some way integrated such faint stimuli as
irregular scattered dots and faint, wavy lines into straight
lines which have no objective existence. Much has
been made of this test, but it may hardly be called conclusive, for Flammarion when repeating the experiment
with French boys was unable to secure in their sketches
lines resembling canals. Moreover, photographs taken
during the opposition of 1905 at Flagstaff bring out a
large number of the more prominent canals as straight
lines.

It is possible that psychology plays an important part,
and Professor Andrew Ellicott Douglass, of the University of Arizona, who has had opportunities to observe
the planet at several favorable oppositions, believes that

the subjective phenomena have much to do with Martian observations. He says:

"One may confidently say that such realities do exist. But with the very faint canals whose numbers reach occasionally well into the hundreds, discordance reigns supreme, and it is frequently found that different drawings by the same artist antagonize each other across the page.

"Considerations along these lines led me to study seriously the origin of these inconsistent faint canals by the methods of experimental psychology, and the application of these methods has resulted in a new optical illusion and new adaptations of old and well-known phenomena." The most important of these phenomena was the halo effect where secondary images are produced under unusual conditions, which images affect the minute details of the surface. Professor Douglass also found that the irregular refraction of the eye produced apparent rays as from a small spot, which could be obviated in part by changing the position of the observer's head. "The ray illusion," he says, "is to me a very satisfactory explanation of many faint canals radiating from those small spots on Mars called 'lakes' or 'oases.' The only objective reality in such cases is the spot from which they start."

Referring again to the halo he reaches the conclusion that—

"The halo with its light area and secondary image accounts for details which have no objective reality, such as bright limbs of definite width, canals paralleling the limb or dark areas, numerous light margins along dark areas and light areas in the midst of dark, abundantly exemplified in Schiaparelli's map of 1881-82."

Professor Lowell, it must be said, has the unique advantage, or misfortune it may be, to see canal-like markings that are not visible to other astronomers. Thus on Venus he saw bands or lines which he considered bore a superficial resemblance to the canals of Mars and were apparently permanent and not due to clouds. Again he claims

to have seen lines resembling canals on the third satellite
of Jupiter, which others have failed to recognise. Hence
many astronomers believe that he is predisposed to see
such phenomena. Furthermore, canal-like appearances have
been noted on the surface of the Moon by Professor W. H.
Pickering as radiating from the central peaks of the north-
western slopes of the central mountain range of the crater
of Eratosthenes. He observed two canals which in a small
telescope appear straight, yet, when seen with a large
glass, present considerable irregularity of structure. Other
and new branches or canals were also seen. In various
parts of the same crater, but especially in the southeastern
and northern portions, numerous small canals and lakes
present themselves. These markings are practically identical in appearance with those seen upon the planet Mars.
They are too small to be well shown on photographs and
seem to be of much more regular structure than the larger
markings, which are also called canals. It is possible that
this difference is due merely to the fact that the larger
markings are better seen. There can be no free water
on the Moon's surface; hence any canals with flowing
water are quite out of the question. Yet the appearance is
vouched for as remarkably similar to that on Mars by an
observer who knows the surfaces of both bodies.

CHAPTER XVI

As Jupiter comes next to Venus in point of brilliancy of the heavenly bodies, it is but natural that it should have been known to the ancients from a remote antiquity and that its discovery or early observation should be lost in a far distant past. The brightness of Jupiter varies with its position, and the relative brightness of the planet at an average conjunction at the nearest and most remote oppositions is respectively as the numbers 10, 27, and 18.

The orbit of the planet is but slightly inclined to the ecliptic, or 1° 19', and the planet itself moves with an orbital velocity of about eight miles a second in the sidereal period of 11.86 years, which is the time of its revolution around the Sun from a star to the same star again, as seen from the Sun. The mean distance of the orbit of the planet from the Sun is 483,000,000 miles. The eccentricity is nearly one-twentieth, the greatest and least distances from the Sun being 504,000,000 and 462,000,000 miles respectively. The average distance of the planet from the Earth at opposition is 390,000,000 miles, while at conjunction it is 576,000,000 miles. The minimum opposition distance, occurring about October 6, amounts to 369,-000,000 miles, while at aphelion in April the distance is greater by about 42,000,000 miles.

Jupiter is larger than all the rest of the planets in the solar system, whether its bulk or its mass be taken into consideration. Its surface is 119 times and its volume

1,300 times that of the Earth. The mean diameter is 86,500 miles, or about eleven times that of the Earth. But if the relative masses and volumes of the two planets be compared, it will be found that Jupiter has a density less than one-quarter that of the Earth, or .24, and almost the same as that of the Sun. A body on Jupiter would weigh 2⅝ times as much as upon the surface of the Earth, because the mean superficial gravity of the planet is 2.64 times as great. Owing to its rapid rotations and its elliptical shape the difference between the force of gravity at the equator and the pole is much greater and amounts to ⅕ of the equatorial gravity, where on the Earth it is as ¹/₁₀₀.

The planet is brightest, as is also Saturn, in the center of the disk, which it will be recalled is the case with the Sun, but not with Mars, Venus and Mercury. On account of this resemblance to the Sun, the idea has been suggested that the planet may be, to some extent, self-luminous.

The planet receives too small an amount of heat from the Sun to account for the rapid changes which beyond question are taking place on its visible surface. Consequently, to produce these changes the heat must come from the planet itself. Probably the body is at a temperature little below that of incandescence, not having solidified to any appreciable extent.

The most striking feature of Jupiter is its system of bright satellites, four of which were the first fruits of astronomical discovery as it is now understood, and were revealed to Galileo when he directed his small telescope toward the planet. In fact, this historic observation of January 7, 1610, meant much to astronomy. When the great Italian scientist determined the periods of these strange bodies, or "Medicean stars," as he termed them, a new era was opened in astronomy. The four satellites, in addition to being numbered in the order of their distance from the planet, are also known by the mythological names of Io, Europa, Ganymede and Callisto, and revolve in sidereal periods, ranging from 1 day, 18½ hours to 16 days, 16½ hours at relative distances of between 262,000 and 1,169,000 miles. From the small telescope of Galileo in the opening years of the seventeenth century to the Lick refractor at the close of the nineteenth is indeed a far cry, but the four satellites of Jupiter remained alone until a fifth was added to their number by Professor E. E. Barnard at Mt. Hamilton in September, 1892. This discovery was as much a triumph for the modern telescope as the original detection of the four moons was an achievement for the "Optick Tube," for the satellite is visible only with telescopes having a greater aperture than 18 or 20 inches. It has a period of 11 hours, 57.4 minutes, and its nearness to the planet, 112,500 miles, makes it additionally difficult to see.

But this was by no means the end, for where the eye failed the photographic plate was available. In December, 1904, and January, 1905, Professor C. D. Perrine of Lick Observatory added, by photography, two new moons to Jupiter's system. Both of these bodies revolve at a greater distance than the older known satellites. Still more recently P. Melotte of Greenwich Observatory, while examining a photograph made there on February 28, 1908, found a faint object which proved to be an eighth satellite of Jupiter, photographed several times at Greenwich, at Heidelberg and at Lick Observatory. The movement is retrograde, which anomaly is of vast cosmical importance.

The discovery of the satellites of Jupiter by Galileo was still another point which brought him in conflict with the Church. In 1611 there was published a tract in which it was mentioned that the satellites of Jupiter were unscriptural. This, apparently, was a minor issue with the ecclesiastical authorities, for the evidence of the telescope was incontrovertible.

Galileo believed that it was possible to determine the longitude at sea by means of the satellites of Jupiter, and corresponded with the Spanish Court in reference to a method which he had devised. He held that if the movements of Jupiter's satellites and, in particular, the eclipses which constantly occur when the satellites pass into Jupiter's shadow, could be predicted, then a table could be prepared giving the dates at some standard place, say Rome, at which the eclipses would occur. The local time of the eclipse could be readily observed and referred to the local time—that is, its noon or when the Sun is highest in the sky, with no great amount of error, and the difference in time between the two places would naturally give the difference in longitude.

In Galileo's day astronomers and navigators had no accurate means of keeping time, such as the modern chronometer, which, carried on a ship, can be kept at Greenwich or some other standard time and give the difference between that and local time immediately. Galileo knew nothing of the pendulum clock of Huygens, or, more especially, the chronometer of John Harrison (1693-1776), which has made possible the accurate determination of longitude at sea. The motions of the satellites continued to arouse general interest, and Kepler, taking the movements of the four satellites around the parent planet, as recorded by Galileo and Simon Marius, found that his laws of planetary motion applied to the satellites as well as to the planets themselves.

But perhaps the most striking of the discoveries made with Jupiter, after the actual detection of its satellites, was that of the Danish astronomer, Olaus Roemer (1644-1710), in 1675, when engaged in the study of the motion of Jupiter's satellites. He ascertained that the intervals between successive eclipses of a moon, which were caused by its passage into Jupiter's shadow, were regularly less when Jupiter and the Earth were approaching each other than when they were receding. Accordingly he made the ingenious assumption that light travels through space, not instantaneously, but at a certain definite tho very great speed.

Accordingly, if Jupiter is approaching the Earth, the time which the light from one of his moons takes to reach this planet must be gradually decreasing, and consequently there is less interval between successive eclipses as seen from the Earth than when the great planet is departing from it. Now the difference of the intervals thus observed, together with the known rates of motion of Jupiter and of the Earth, which, of course, could be calculated, made it possible to form a rough estimate of the speed at which light travels. Roemer had not sufficient observations at his command to investigate this problem very thoroly, but he was able to compute the apparent retardation of the eclipses between opposition and conjunction and thus to obtain a value for twice the time required for light to come from the Sun to the Earth, which time was very nearly 500 seconds, or eight minutes and twenty seconds. This was the first work on the so-called "equation of light." It was many years before astronomers accepted Roemer's really wonderful method for obtaining the distance from the Sun. To-day the process is reversed. By elaborate physical experiments made on the Earth's surface it is possible to obtain an accurate value for the velocity of light and then by means of the light equation to deduce the distance from the Sun.

James Bradley (1693-1762), the third Astronomer Royal of Great Britain, also devoted himself to the study of the satellites of Jupiter. With Cassini's observations, which he used as the basis for some tables, as well as many of his own dealing with the eclipses of the satellites, he noted a large number of discrepancies between the observations and the tables, and found even more peculiarities in their motions than did the early observers. Using Roemer's suggestion of the finite time consumed by light in traveling from Jupiter to the Earth, which Cassini and other astronomers of his time had rejected, Bradley was able to make a series of new and valuable tables of Jupiter's satellites, which were printed in 1719 in Halley's "Planetary and Lunar Tables." Bradley's knowledge of the satellites of the planet was applied to the method of the determination of longitude suggested by Galileo, and with great accuracy he found the longitudes of Lisbon and of New York.

Jupiter rotates on its axis, which is inclined about 3° to the orbit, once in about 9 hours and 55 minutes, a time which is difficult to obtain more than approximately, for when different spots are observed different results are obtained. These spots were first observed in 1665 by Giovanni Domenico Cassini (1625-1712), who was also the first to study the so-called belts. He was able to report the discovery of the rotation of the planet by watching the movement of the spots when observed through the telescope. One of these spots is the famous "great red spot" first observed in modern times by Professor C. W. Pritchett in Glasgow, Missouri, in July, 1878.

This is a rosy cloud attached to the whitish zone beneath the dark southern equatorial band. Of enormous size, measuring some 30,000 miles in longitude and somewhat less than 7,000 miles in latitude, it was seen by several observers in Europe in the year of its discovery, and in the following year was observed by almost every astronomer possessed of a telescope. For three years the red spot was conspicuous. Then it began to fade. When the planet returned to opposition in 1882 and 1883 Rica's observations of it at Palermo, May 31 1883, were expected to be the last. It began to recover, however, toward the end of the year, and at the beginning of 1886, according to W. F. Denning, an English observer, had much the same aspect as in October, 1882.

Before the "great red spot" astronomers had noticed various markings on the planet, one of which, as we have seen, was recorded by Cassini in 1665 as having a rotation period of 9 hours and 56 minutes. This spot reappeared and vanished eight times within the next forty-three years and was last seen by Maralda in 1713. It was, however, very much smaller than the recent object and showed no unusual color. Agnes M. Clerke, from whose 'History of Astronomy' is abstracted this brief description of the "great red spot," further discusses the phenomenon as follows: "The assiduous observations made on the 'Great Red Spot' by Mr. Denning at Bristol and by Professor Hough at Chicago afforded ground only for negative conclusions as to its nature. It certainly did not represent the outpourings of a Jovian volcano; it was in no sense attached to the Jovian soil—if the phrase have any application to that planet; it was not a mere disclosure of a glowing mass elsewhere seethed over by rolling vapors. It was, indeed, certainly not self-luminous, a satellite projected upon it in transit having been seen to show as bright as upon the dusky equatorial bands.

"A fundamental objection to all three hypotheses is that the rotation of the spot was variable. It did not then ride at anchor, but floated free. Some held that its surface was depressed below the average cloud level and that the cavity was filled with vapors. Professor Wilson, on the other hand, observing with the 16-inch equatorial of the Goodsell Observatory in Minnesota, received a persistent impression of the object's 'being at a higher level than the other markings.'

"A crucial experiment on this point was proposed by Mr. Stanley Williams in 1890. A dark spot moving faster along the same parallel was timed to overtake the red spot toward the end of July. An unique opportunity hence appeared to be at hand for determining the relative vertical depths of the two formations, one of which must inevitably, it was thought, pass above the other. No forecast included a third alternative, which was nevertheless adopted by the dark spot. It evaded the obstacle in its path by skirting around the southern edge.

"Nothing, then, was gained by the conjunction beyond an additional proof of the singular repellent influence exerted by the red spot over the markings in its vicinity.

It has, for example, gradually carved out a deep bay for its accommodation in the gray belt just north of it. The effect was not at first steadily present. A premonitory excavation was drawn by Schwabe at Dessau, September 5, 1831, and again by Trouvelot, Barnard and Elvins in 1879; yet there was no sign of it in the following year.

"Its development can be traced in Dr. Beddicker's beautiful delineations of Jupiter, made with the Parsonstown 3-foot reflector, from 1881 to 1886. They record the belt as straight in 1881, but as strongly indented from January, 1883; and the cavity now promises to outlast the spot. So long as it survives, however, the forces at work in the spot can have lost little of their activity. For it must be remembered that the belt has a shorter rotation-period than the red spot, which, accordingly (as Mr. Elvins of Toronto has pointed out), breasts and diverts, by its interior energy, a current of flowing matter, ever ready to fill up its natural bed and override the barrier of obstruction."

The object is now always inconspicuous and often practically invisible, and may be said to float passively in the environing medium. Yet there are sparks beneath the ashes. A rosy tinge faintly suffused it in April, 1900, and its absolute end may still be remote.

Besides the spots, Jupiter exhibits curious belts or bands. Herschel, that observer 'par excellence,' frequently turned his telescope to Jupiter as to other planets, and became greatly interested in its bright bands. In 1793 he was the first to interpret these as bands of clouds. In fact, telescopic examination of Jupiter during the nineteenth century established the fact that the visible surface of the planet appears as layers of clouds, and its low density, 1.3, as compared with water, 1, and the Earth, 5.5, together with the rapid changes, indicates that the planet is, to a great extent, in a fluid condition, and that there is a high temperature at a very moderate distance below the visible surface.

CHAPTER XVII

THE planet Saturn was considered by the ancients to be the most distant of the moving heavenly bodies, a position it retained even after the triumph of the Copernican ideas and the establishment of the modern conception of the solar system. The reason for this was that the period of its oscillatory motion to and fro in the heavens was longest of all of the planets. The ancients noted that it took 29½ years for Saturn to return to the same place among the stars, as compared with 12 years for Jupiter, and correspondingly less for the other planets down to 88 days for Mercury. Accordingly they considered that Saturn was the most distant.

Eudoxus (409 B.C.[?]-356 B.C.[?]), as has been shown, believed that the motion of Saturn, as in the case of Mars and Jupiter, could be represented roughly by supposing that each planet oscillated to and fro on each side of a fictitious planet which moved uniformly around the celestial sphere in or near the ecliptic. The slow period of Saturn also made it the most distant of the planets in the system as devised by Copernicus, who computed its distance from the Sun as nine times that of the Earth, which may be compared to his credit with the modern figure of 9½ times.

After the development of observational astronomy Tycho Brahe in 1563 made his first recorded observation at the University of Leipzig, noting the close approach of

Jupiter and Saturn, which he found was quite a month in error in the prediction of the Alfonsine Tables, published in Spain in 1252 and in general use by astronomers' throughout Europe. The next important observation of Saturn was indeed of epoch-making significance in astronomy. With his new telescope it was but natural that Galileo should examine Saturn as he did the other planets. Turning his telescope toward Saturn he observed that that planet, too, was not single and complete, but apparently consisted of three parts, or, as it appeared in a drawing made at the time by him, of a central body and two satellites in close proximity, which naturally seemed to resemble those of Jupiter. At subsequent observations he failed to see more than the central and larger portion, and, consequently, completely baffled, he left the problem as a legacy to his successors.

The two appendages were seen and described under varying conditions by a number of astronomers, but the true solution was first furnished by Huygens (1629-1695), when he studied with one of his powerful telescopes the appearance of the planet. Huygens announced in 1655 that he had discovered a single satellite, which he named Titan. With a still more powerful instrument he found that the effect of two component bodies observed by Galileo was due to the fine ring which surrounded the planet and was inclined at a considerable angle to the plane of the ecliptic and consequently to the plane in which Saturn proceeds around the Sun. As the ring was extremely fine it became invisible, either when its area was directly opposite to the observer or when it was directed toward the Sun, as in that case it received no light for reflection. Near this opposition or invisibility the ring appears to be foreshortened and presents the appearance of two arms projecting from the body of the planet. The ring, of course, gradually widens from its opposition or invisibility and the opening becomes visible, a period of seven years e.asping between such a state and when the ring is seen at its widest.

With the observations of Huygens the reasons for Galileo's varied observations were furnished. To make the matter more conclusive Huygens collected and published a series of early drawings by various observers, which drawings he compared with his own observations. Thus what Galileo conceived as two satellites was really the

Fig. 25 —Relative Sizes of Saturn and the Earth.

ring when seen with its greatest breadth. The disappearance of these satellites occurred when the edge of the ring was presented to his view, the revolution of the planet giving to an observer on the Earth a series of phases in which the appearance of the planet is remarkably different.

What Huygens saw is now familiar to every one who has observed Saturn through a telescope. Surrounding the central body are rings parallel to the planet's equator, but inclined about 27 degrees to the plane of its orbit and

28 degrees to the ecliptic, their nodes being at longitude
168° in Aquarius and at longitude 348° in Leo. The plane
of the rings remains sensibly parallel to itself for a very
long time. For fifteen years, or half a revolution of Saturn, their northern face is seen and during the remaining
half of the revolution their southern face. When the
Earth passes over the plane of the rings at the time of
transition their edge is presented so that the ring virtually disappears from view, as occurred, for example, in
1908, and in 1612 had occasioned Galileo's perplexity.
The thickness of the rings is less than 100 miles; consequently their edge was quite invisible through his feeble
telescope. The disappearance of the rings recurs at intervals of about fifteen years.

In 1675 Giovanni Domenico Cassini (1625-1712) noticed a dark marking in the ring which later was found
to mark the division of the ring into two distinct schemes,
a narrow and outer ring, to which the name of "Cassini's
Division" was given. As was natural, the peculiar construction of the rings of this planet and the accompanying satellites was the subject of deep and earnest inquiry,
both mathematical and telescopic, tho but little substantial
progress was made in explaining the formation and occurrence. In his analysis of the nature and motions of the
planets and their satellites the mechanical problem of the
stability of Saturn's rings was left even by Laplace in a
very unsatisfactory condition; for he made no attempt
to determine the kind or amount of irregularity in the distribution of their weight, which he assumed was necessary
in any considerations of them as rotating solid bodies.

In 1849 W. C. Bond at Cambridge, Mass., found that
Saturn was accompanied by a third comparatively dark
ring, lying immediately within the bright rings, and to
this the name "Crape Ring" has been applied. Professor
Bond, who devoted much attention to the study of Saturn,
at first denied the solidity of the planet's structure and
asserted that the fluctuations in its aspect were entirely

at variance with any such hypothesis. He and other astronomers had frequently detected in both of the bright rings fine dark lines of division, and as these frequently lapsed into imperceptibility the condition was due in his opinion to the real mobility of their particles and indicated a fluid formation. The known solidity of the rings was then demonstrated on abstract grounds by Professor Benjamin Peirce of Harvard University, who maintained that they were formed by streams of some fluid denser than water. In 1857, in England, James Clerk Maxwell, the famous mathematician and physicist, presented a mathematical discussion of the subject, in which he stated that neither solid nor fluid rings could exist and that the system could be composed only of a great multitude of uncollected particles which revolved independently in a period corresponding with their distance from the planet. This idea of a satellite formation, remarkably enough, had been several times entertained and lost sight of, so that when advanced by Maxwell it was a virtual novelty. The hypothesis met the test of telescopic observation.

The mathematical theory of the ring system found an analogy in the assemblage of planetoids, both visible and invisible, which are known to be revolving around the Sun with orbits situated between Mars and Jupiter. If seen from a considerable distance such a swarm of these small particles would give the impression of a continuous solid body, so that on its physical basis this theory did not seem improbable. It was pointed out by Kirkwood in 1867 that the division between the two main orbits, first made by Cassini, could be explained by the perturbations due to certain of the satellites, just as the corresponding gaps of the minor planets are explained by the action of Jupiter. But in modern astronomy the probability of a mathematical theory is not sufficient, and its acceptance must depend upon direct and conclusive evidence from telescope, spectroscope or photographic plate.

This was supplied most effectively by Professor James E.

Keeler (1857-1900), at Allegheny Observatory, Pittsburg, in 1895. He pointed his spectroscope to the planet and found by examining the light waves from opposite sides that the main body was in rotation. The light from one side was approaching the Earth and from the other it was receding. This rotation of the planet, of course, had been realized for some years, but the axial rotation of the rings had never before been demonstrated, and this Keeler proved, revealing the strange fact that the interior part of the rings rotated faster than the exterior, which of course would not hold true in the case of a solid body. The motion slowed off outward in agreement with the diminishing speed of particles traveling freely, each in its own orbit.

The visibility of the rings when the Sun and the Earth are on opposite sides of their plane is explained by Professor Barnard as due to the filtration of sunlight through a cloud of cosmical dust. He regards the knots as the result of the radiation of parts of the clouds which are denser, but not necessarily thicker than the rest under the illumination of sunlight, which gives to them their adjacent portions of less density. Percival Lowell, however, believes that the rings of Saturn are not flat and of uniform thickness, but rather resemble a concentric series of tores or anchor rings, and that the knots represent their fixed portions.

Professor Sir G. H. Darwin explains the rings of Saturn by considering them an abortive satellite, scattered by tidal action into annular form, for they lie closer to the planet than is consistent with the integrity of a revolving body of reciprocal bulk. This interesting appendage, according to Professor Darwin, will eventually disappear, as the constituent particles will be dispersed inward in part and will be gathered to the surface of the planet, while in part they will scatter outward where they may coalesce unhindered by the strain of unequal attraction. Then one modest planet revolving within Mimas would be all that would remain of appurtenances which lend character to the planet.

The dimensions of these wonderful rings of Saturn doubtless will arouse the reader's curiosity. The planet itself has an equatorial diameter of 75,000 miles. Outside of this first comes the crape ring at a distance of from nine to ten thousand miles of clear space, and somewhat less than 10,000 miles in width. The crape ring is joined to the second ring, which is the most brilliant and is about 16,500 miles in width. Then comes a gap of about 1,600 miles, and there lies the outer ring, 10,000 miles in width, with an exterior diameter of about 168,000 miles. Hence the entire ring system has a width of between 36,000 and 37,000 miles. A model of the outer ring, constructed on the scale of 10,000 miles to the inch, could be made with an approximation to accuracy from a sheet of writing paper nearly seventeen inches in diameter.

Cassini's discovery of the dark markings in Saturn's ring was one result of a series of telescopic observations which he made of the planet, in which he discovered four new satellites—Japetus in 1671, Rhea in 1672, and Dione and Thetis in 1684. This list of satellites was increased by two more in 1789, when Herschel, using his 40-foot telescope for the first time, August 28th, detected a sixth satellite of Saturn, Enceladus, and on September 17th discovered a fainter satellite, Mimas. Both of these were nearer to the planet than any of the five previously observed. In September, 1848, W. C. Bond, of Cambridge, Mass., discovered Hyperion, and two days later this same satellite was also observed at Liverpool by William Lassell.

A ninth satellite of Saturn, Phœbe, was discovered by Professor W. H. Pickering on photographic plates taken at the Harvard Observatory at Arequipa, Peru, and was announced in July, 1904, the satellite being seen on a number of recent photographs. This satellite is much smaller than any of the existing moons; so much so, in fact, that it is beyond the visibility of the human eye with any existing telescope. It revolves around Saturn at a distance of many millions of miles, far beyond the orbit of Japetus and with a period correspondingly longer, and strange to say, in an opposite direction from its fellows.

Professor Pickering also discovered in 1905 a tenth satellite of Saturn, Themis, which revolves much closer to the planet. It is said to be the faintest object in the solar system, and is a striking illustration of what astronomical photography can accomplish in the way of discovery.

Saturn moves in an orbit which is somewhat more eccentric than that of Jupiter, but which is at a mean distance from the Sun of 886,000,000 miles. The equatorial diameter is about 75,000 miles and the polar diameter about 68,000, giving a mean diameter of 73,000 miles, or a little more than nine times that of the Earth and a volume greater by 760 times. Yet Saturn is a very light body, having a mass only 95 times that of the Earth and a density of one-eighth, which would give it a specific gravity of five-sevenths that of our own planet.

The same arguments advanced in favor of a high temperature for Jupiter can be used with increased force in the case of Saturn. It may be assumed that a large proportion of this bulky globe is composed of heated vapors which are vigorously circulated by the process of cooling.

Professor Asaph Hall, of Washington, in 1876 made observations of the white spot which was visible on the surface of the planet for some weeks. He established a period of rotation of 10 hours, 14 min. and 24 sec., which agrees closely with 10 hours, 16 min. determined by Herschel in 1794. Hall's value has been confirmed by other observers and is generally accepted.

Saturn shows belts similar to those of Jupiter, with a brilliant zone at the equator. The edges of the disk are not so brilliant as the central portion, for the pole of the planet is at times marked with a darkish cup of greenish color.

CHAPTER XVIII

URANUS AND NEPTUNE

THE discovery of Uranus distinctly represents one of the most important results of modern methods in astronomy. The other planets considered were known from prehistoric times. Even the least conspicuous of them could be observed with the naked eye under favorable conditions. Just as the satellites of Jupiter were the first fruits of telescopic discovery in the heavens, so Uranus was the first planet to be telescopically added to the list which through centuries had been known and studied.

Its discovery was made on March 13, 1781. by Sir William Herschel while engaged in the systematic examination of every stellar body visible with his 7-foot reflecting telescope. Herschel states that: "In examining the small stars in the neighborhood of 'H. Geminorum' I perceived one that appeared visibly larger than the rest; being struck with its uncommon appearance I compared it to 'H. Geminorum' and the small star in the quartile between 'Auriga' and 'Gemini,' and finding it so much larger than either of them I suspected it to be a comet."

Though it appeared as a star of the sixth magnitude, its difference from the other stars was at once appreciated by Herschel and is evidence of that keenness of sight which was so characteristic of him. Observations of the new body and study of its orbit failed to establish it as a comet. Within three or four months of its discovery the conclusion was reached first by Anders Johann Lexell

(1740-1784) that it was a new planet which revolved around the Sun in an orbit nearly circular at a distance of about nineteen times that of the Earth and nearly double that of Saturn.

Herschel's discovery at once won for him a national reputation and royal honors, which he attempted to reciprocate by conferring the name of his royal patron, George III., on the new planet, calling it Georgium Sidus. But the name never gained currency outside of England. After a vain attempt to apply Herschel's name to it, the old mythological nomenclature was observed and the new planet became permanently known as Uranus, at the suggestion of Bode, after the father of Saturn and the grandfather of Jupiter. The name means Heaven itself, beyond which it was supposed nothing further could be found.

Following its discovery by Herschel, with a reflecting telescope 40 feet in length and of 4-foot aperture, came the detection on January 11, 1787, of two moons or satellites of Uranus, to which the names of Oberon and Titania were subsequently given. Herschel discovered that these moons moved almost at right angles to the ecliptic in a direction contrary to that of all previously known members of the solar kingdom except the comets. He suspected the existence of four more such satellites, but he was not able to assure himself positively of their existence. In fact, it was only his large telescope and his keen eye that enabled the first two moons to be observed. But with the progress of astronomy and the improvement of instruments other discoveries were bound to come.

Within the paths of Oberon and Titania, Ariel and Umbriel were found, October 24, 1851, by William Lassell, a wealthy brewer, who during his life devoted himself assiduously and with great success to astronomy, especially telescopic observation. These satellites, altho not easily visible in a telescope on account of their distance, are much larger than the satellites of Mars or in fact many of the planetoids. It is estimated that their diameters are between 500 and 1,000 miles. Oberon, which is distant from the planet 365,000 miles, has a period of rotation of 13½ days; Titania, the largest and brightest, distant 273,000 miles, has a period of 8.7 days; Umbriel, distant 167.000 miles, has a period of 4.1 days; and Ariel, distant 120,000 miles, has a period of 2.5 days. These satellites all move in the same plane, which is inclined about 98° to the plane of the planet's orbit. The satellites revolve in a retrograde direction.

The surface markings observed on Uranus by many astronomers have been vague and transitory, so that any determination of the period of rotation is but approximate. Nevertheless, a period of 10 or 12 hours has been indicated. It is stated that the plane of the equator is inclined something like 10 to 30 degrees to the plane of the orbit of the satellite. The disk of the planet shows a flattening at the poles, so that it has an elliptical section.

The appearance of Uranus to the naked eye is that of a small star of about the sixth magnitude. It was on this account that a high power telescope was required to differentiate it from the myriad other stars of this size to establish it as a planet. It is so far away that there is but little change in its position whether it is in opposition or quadrature. Measuring the disc, which appears in the telescope to be of a sea-green color, the diameter of the planet is found to be about 32,000 miles, or four times as great as that of the Earth, which would give it a volume 64 times greater. But, like the other distant planets, Uranus is composed of lighter materials, so that while 64 times as large its mass is but 15 times that of the Earth, or, in other words, Uranus would compare with the Earth in about the same proportion as regards volumes as does the Moon with the Earth.

The elliptical orbit in which Uranus moves at a mean distance from the Sun of nearly 1,800,000,000 miles requires 84 years for its passage, and the diameter of this orbit is 3,600,000,000 miles. The orbit is slightly less eccentric than that of Jupiter and amounts to 83,000,000 miles, while the periodic time of the planet is 84 years and its synodic period 369 days and 16 hours. It moves with an orbital velocity of 4⅓ miles per second. In the first half century after its discovery Uranus gave astronomers considerable trouble, as observations showed that it was not following exactly its computed path and that it deviated by a substantial amount. About 1830 Bessel suggested that the discrepancies in the observed and calculated orbits might be due to an unknown planet, then more distant from the Sun than Uranus, and such was found to be the case.

As Uranus was the triumph of telescopic discovery, so Neptune represents one of the greatest achievements of mathematical astronomy. In fact, when the French astronomer, Leverrier, at Paris, wrote to Galle, at Berlin, substantially as follows, "Direct your telescope to a point on the ecliptic in the constellation of Aquarius in longiture 326° and you will find within a degree of that place a new planet looking like a star of about the ninth magnitude, and having a perceptible disk," the German astronomer, within thirty minutes after he had begun his search, on the night of September 23, 1846, was able to find the new planet but 52' distant from the point indicated by Leverrier.

The discovery of Neptune came, as has been suggested, from discrepancies observed in the path of Uranus, which oftentimes were almost so marked as to be observed without the aid of the telescope. As an explanation of the disturbance, an unknown exterior body was suggested, which was not only plausible but so obvious that several astronomers were devising mathematical plans of campaign for its discovery.

A young graduate of Cambridge University, John Couch Adams (1819-1892), who had distinguished himself in mathematical work, assiduously addressed himself to the problem, and in 1845 sent to the Astronomer Royal at Greenwich numerical estimates of the elements and mass of the unknown planet, together with an indication of its actual place in the heavens. Unfortunately, Adams' work, for various reasons, was not taken up by the government astronomers, and in the meantime Urbain Jean Joseph Leverrier (1811-1877), as a result of the study of the stability of the solar system, and especially of the Uranian difficulty, to which his attention had been directed in 1845 by Arago, announced before the French Academy that only an exterior planet could produce the observed effects.

Such an announcement aroused astronomers to a point of expectancy, and in fact, as Sir John Herschel declared to the British Association regarding the hypothetical new planet, "We see it as Columbus saw America from the coast of Spain." In less than two weeks from the time of this utterance the message quoted was sent to Galle at Berlin. Within a week Neptune was also observed in England, where delays lost the honor of priority for Adams.

Once the existence of the new planet was established, a few weeks' observation made possible the computation of its orbit and its identification with what had been considered a fixed star.

Neptune supplied the exception which proved the rule in the case of Bode's law, discussed in the chapter on planetoids, for its mean distance from the Sun was 2,800,-000,000 miles instead of 3,600,000,000 as would be required under the terms of the law. Furthermore, Neptune has an orbit with an eccentricity of only $^9/_{1000}$, so that its path is more nearly circular than that of any other member of the solar system except Mercury. But so large is its orbit that the small eccentricity makes a variation of over 50,000,000 miles in the distance of the planet from the Sun at different parts of its orbit. Moving with an orbital velocity of about $3\frac{1}{3}$ miles a second, it requires 164 years for its journey around the Sun.

Neptune has but one satellite, which Lassell found in 1846, within a month of the original discovery of the planet. Its distance is about 221,500 miles and its period of revolution is 5 days and 21 hours. It is a small body, about 2,000 miles in diameter, or about the size of the Earth's Moon, and it moves backward just as the satellites of Uranus, in an orbit that is inclined 145° to that of the planet.

Like Uranus, Neptune varies little, as its distance is so great that any variation by change in the position of the planet would not affect its appearance on the Earth. Its diameter is estimated at about 35,000 miles, which would give a volume 85 times that of the Earth (or according to Professor T. J. J. See, U. S. N., 27,190 miles); but, as also the case with Uranus, it is much lighter than the Earth. As its density is probably about 0.20, its mass, which astronomers compute from the motion of its satellite, is about seventeen times that of the Earth. On account of its great distance it receives from the Sun about $\frac{1}{906}$ the amount of heat that falls upon the Earth. If its capacity for absorbing and retaining heat are the same, the theoretical temperature would be about 360° F. (220° C.), or between the temperature of liquid air and the boiling point of hydrogen, a temperature so cold that on the Earth complex methods to produce it have to be adopted in a physical laboratory. Nevertheless, the amount of light received from the Sun is not insignificant, and the noonday illumination of the planet would be some 700 times that of brightest moonlight. If the Sun were placed at Neptune, its light would equal that of 687 full moons.

No surface markings have yet been seen on Neptune. Consequently astronomers are unable to determine its rate of rotation by direct observation; but from various intricate processes Neptune is believed to have a slower rotation than Jupiter or Saturn.

CHAPTER XIX

THE PLANETOIDS

In the chapter on the Solar System it has been shown how Titius in 1772 pointed out that if 4 should be added to the following series (irregular in that the first number should be ½ instead of 0): 0, 3, 6, 12, 24, 48, a sequence of numbers would be obtained that indicated the relative distances of the six planets, with the striking exception that next after Mars there was no representative. Bode filled this gap with a hypothetical planet, and his name is often associated with that of Titius in the statement of this law.

After the discovery of Uranus by Herschel in 1781 the law of Titius or Bode, with which this planet conformed, gained considerable respect from astronomers, altho no mathematical or other reason was known to uphold the singular relation between the distances of the planets from the Sun. At any rate, the conviction that there must be an undiscovered planet between the orbits of Mars and Jupiter, as was even suspected by Kepler in his "Mysterium Cosmographicum" of 1596, was strengthened, and accordingly it was proposed by a voluntary association of astronomers in 1789 to undertake a systematic search for the missing planet. An organization was arranged on a fairly methodical basis by Baron Von Zach; but before these "celestial police," as the astronomers enlisted for this purpose were humorously termed by their organizer, had secured results from their plan of coöperation the missing planet was found under interesting and somewhat extraordinary circumstances.

At an observatory at Palermo, Sicily, Giuseppe Piazzi (1746-1826) had been at work for some nine years on the preparation of a catalogue of the stars. On January 1, 1801, he noted an eighth magnitude star, which on subsequent evenings shifted its position and induced him to believe that he had discovered a new kind of comet without tail or coma, and so he described it in a letter to Bode at Berlin. A fortnight later the wandering body changed its retrograde for direct motion, and before the observations could be repeated by other European astronomers Piazzi's moving star approached too near the Sun to be visible any longer. For its rediscovery some accurate knowledge of its path was essential.

Never before had there been such meager data on which to calculate the motion of a celestial body, for even in the case of Uranus observations for almost a century had been made and recorded on the assumption that that planet was a fixed star. That the new star was a planet, however, lay outside the realm of consideration. It soon was found that the calculation of its orbit could not be adjusted, for the observations did not harmonize with any sort of parabolic cometary orbit. Then the supposition arose that it was a planet and that it had an elliptical orbit. But how was such a path to be calculated? The case had never yet arisen where a planet which had a relatively rapid motion had been under observation for so short a time and the orbital velocity of which was unknown. Men of the hour were not wanting, and a brilliant young German mathematician, Carl Friedrich Gauss (1777-1855), of Göttingen, now entered the ranks of astronomy. He computed the orbit of the new body with remarkable skill, and in November of the same year presented his conclusions to the observing astronomers. It was only on the last night of the year that the sky was sufficiently clear for good seeing. At the Gotha Observatory, almost in the very position assigned by Gauss to the runaway planet, a strange star was discovered, which filled the gap in the series of Titius, and in the following year (March, 1802), at Bremen, Dr. Heinrich Olbers (1758-1840) also observed a similar star. The first of these, at the request of Piazzi,

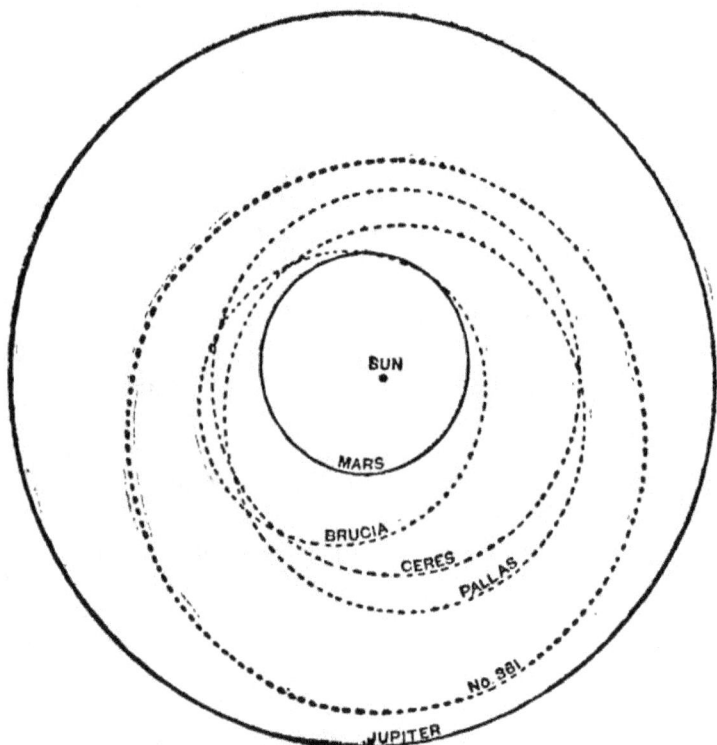

Fig. 26 —Paths of Minor Planets.

was named Ceres, after the protecting divinity of Sicily, while the second, which furnished an additional problem for Gauss, was named Pallas. Harding, the assistant of Schröter at Lilienthal, discovered (1804) Juno as the third, and Olbers (1807) Vesta, the brightest of all the planetoids, as the fourth—all lying between the orbits of

Mars and Jupiter. There were thus, in contradiction to the law of Titius, instead of one planet, four present in the zone between Mars and Jupiter. They were designated "asteroids" by Sir William Herschel, tho the more modern term "planetoids" is to be preferred.

Thus was inaugurated the discovery of minor planets, which continued during the nineteenth century and is still in progress. Where one new planet might be anticipated according to older doctrines, many were found. Various theories have been advanced to explain their existence and occurrence. That Ceres and Pallas were parts of an exploded planet was an ingenious theory which received for a time considerable acceptance, but which was disproved by Professor Newcomb. The recognition of the minor planets became of the highest importance in astronomical science, and led to the construction of star maps and to further interest in the observations of the heavens.

The discovery of the fifth planetoid did not occur for a number of years and then fell to the lot of an amateur astronomer, Postmaster Hencke, at Driessen, who was industriously searching with his telescope the quarter of the sky lying opposite the Sun. He made a practice of mapping all the little stars and of comparing them again on the following nights. At last (1845) he found a new planetoid, to which he gave the name Astræa, and in 1847 a second, Hebe. Since then no year has passed without the discovery of one or more new planets. This may be attributed not only to the example of Hencke, with his systematic industry, but also to the perfecting of the telescope and to the publication of printed special star charts upon the basis of the localizations of numerous fixed stars. Thus by the year 1868 the number of little planets had increased to 100; in 1879, 200; 1890, 300; 1895, 400; 1903, 500; 1906, 600; 1907, 700; 1908, 800. Consequently, instead of one looked-for planet, suspected by Kepler, a complete stream is present between Mars and Jupiter, as indicated in Fig. 26.

The zeal of astronomers was directed toward the observation of the numerous newly-discovered heavenly bodies, and the most fortunate discoverers achieving success in this field were Hind (at London), Goldschmidt, an amateur (at Paris); Gasparis (at Naples), Robert Luther (at Bilk near Düsseldorf), Chacornac and the brothers

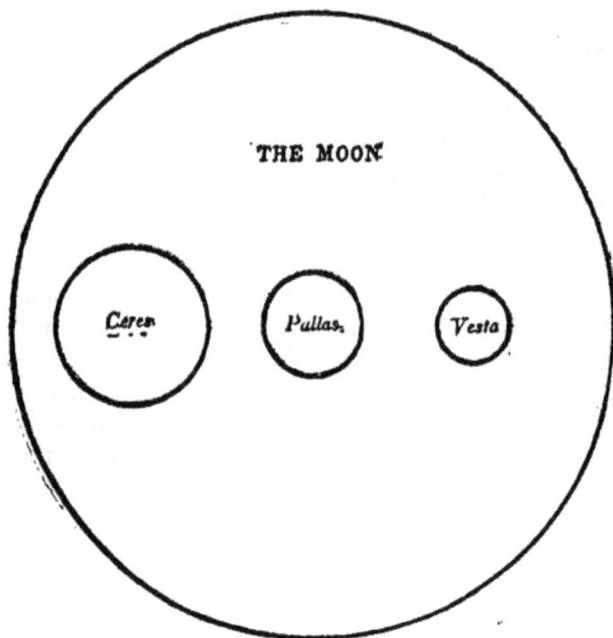

Fig. 27 —Comparative Sizes of the Moon and Three Minor Planets.

Paul and Prosper Henry, who discovered at Paris, during the construction of stellar charts of the zodiac, a number of planets among the numerous charted small fixed stars; and from the United States, Watson at Ann Arbor, and C. H. F. Peters at Clinton, who astonished the world by their numerous discoveries. Johann Palisa should be named above all. While director of the observatory at Pola he discovered 83 new planets, the majority with a small telescope. Later he moved to Vienna, where large refractors were available. The number of small stars now becoming visible changed the accustomed groupings, and thus hindered rather than favored the further discovery of new planets by him. In Vienna, however, he made himself especially serviceable by following up the faint planets already discovered.

Certain orbits of the minor planets are shown in Fig. 26, all of which are described in a standard or west-to-east direction. At the Lick Observatory (Mt. Hamilton, Cal.) in 1894 and 1895 Professor E. E. Barnard made a series of direct measurements of the three largest of the minor planets, obtaining for the diameter of Ceres and Vesta 405 miles for each, and for Pallas 322 miles. The size of these minor planets compared with the Moon is shown in Fig 27. The planetoids in the aggregate do not bulk very large, for we know from the calculations of the distinguished Leverrier, who studied the perturbations of Mars, that the total mass of all known or unknown bodies between Mars and Jupiter cannot exceed a fourth that of the Earth. Later knowledge derived from light observations would place the total mass of those already known at many hundred times less than this limit. If the combined mass were as great as $^1/_{100}$ the mass of the Earth, it would produce perturbations in Mars of which there is no evidence. B. M. Roszel in 1895 estimated that the terrestrial globe contains 3,240 times the aggregate mass of the first 311 minor planets.

In December, 1891, Max Wolff at Heidelberg began that photographic observation and discovery of the little planets, which has resulted in rich gains. Repeated and long exposures of the photographic plate are made, an equatorial telescope fitted with a camera being directed at some point in the heavens, and kept there by means of clockwork. Consequently the fixed stars register themselves upon the plate as points, and the moving planets as short faint streaks, which are discovered by means of the microscope after development and fixing. In this way more planets have been discovered in Heidelberg since 1892 than by all previous observers together. Charlois at Nice emulated Professor Wolff in applying the photographic method, so that there has developed since 1892, with great success, a new field of astronomical photography.

An American astronomer, the Rev. Joel Metcalf, at Taunton, Mass., has improved the technique of observation. In 1905 he found two and in 1906 twelve new planetoids. Since 1892 newly discovered planets have received provisional designations of the letters A, B to Z; AA, AB to AZ; BA, etc., to ZZ, and with the number of the year preceding. If it has once been established that they can be observed sufficiently long to guarantee finding them again, and that they are identical with none of the earlier planets, they receive the next numeral and name, and are thus included in the system. In this manner an unnecessary increase in the lost planets of the system is avoided. In 1907 the crucial moment came in which the planet ZZ was discovered, and the series of provisional designations began again.

Up to 1896 all the 432 planetoids moved in the zone between the solar orbits of Mars and Jupiter; and, indeed, all with the same right-handed motion. Consequently no little attention was attracted when on August 13, 1898, Witt, at the Urania at Berlin, photographically discovered a planet with an unusually rapid motion. The computation of its orbit showed great proximity to the Earth, and a mean solar distance of 1.46 times the distance of the Earth; hence less than that of Mars (1.52). Was this celestial body to be counted in with the stream of planetoids between Mars and Jupiter? Since the possibility existed that still other planets might be discovered which were not confined to the zone between the orbits of Mars and Jupiter, it was determined to include it with the planetoids with the numeral 433, but to distinguish it by the masculine name Eros.

The orbit of this planetoid was found to lie in that forbidden territory within the path of Mars and between it and the Earth. With the single exception of the Moon it is the nearest large heavenly body to the Earth. This little planet, hardly more than twenty miles in diameter, in addition to its anomalous position, possesses remarkable characteristics, among them an orbit of great eccentricity, which causes it to be at aphelion some 24,000,000 miles beyond the mean distance of Mars. An examination of Fig. 28 will reveal how very interesting its orbit is, for on the rare occasions on which the planetoid comes nearest to the Earth it is closer to the Earth than Mars or Venus can ever be.

Furthermore, this little planet not only is variable but varies in its variability, which has given rise to the assumption that either rapid changes are taking place on it, or its light reflecting power varies with its position as regards the Earth and Sun, as might be the case if the planetoid were really double, with two components revolving around a center of gravity, or if the body itself were unequally reflective in its various parts. The mean distance of Eros from the Sun is 135,000,000 miles, but at aphelion it is 165,000,000 miles away. Its perihelion distance is only 105,300,000 miles, or about 12,400,000 miles greater than the mean distance of the Earth. On account of the large inclination of the plane of its orbit to the plane of the ecliptic Eros never approaches the Earth nearer than about 13,500,000 miles.

These near approaches occur only when Eros is in opposition at its perihelion, which unfortunately happens rarely. Its sidereal period is 1.76 years, from which it follows that its synodic period is 2.32 years. If an opposition occurs at perihelion, then in 37 years another will occur very nearly at perihelion, for 37 is almost evenly divisible by 1.76 and 2.32. The next most favorable opposition will occur in 1931, and will furnish an unexcelled opportunity for obtaining the solar parallax.

Because of its eccentric orbit Eros acquires a highly practical significance in respect of the determination of the true size of the entire planetary system. From the laws of the orbits is found, in fact, very accurately (to six decimal places) the relations of the planetary distances. On the other hand, these distances themselves are very little known. They are inversely proportional to the solar parallax (8.80 sec.), of which one can be confident of but

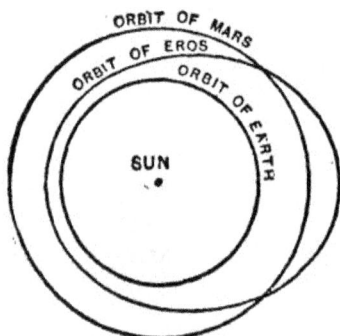

Fig. 28 —The Orbit of Eros Referred to Those of the Earth and Mars.

two decimal places. We have already seen that by means of observation of the transits of Venus, observation of the nearer planetoids and of Mars, determinations of the velocity of light and of the parallactic lunar equation, great pains were taken in the last century to ascertain with more exactness the solar parallax, and thus the extent of our entire planetary system. It appears that the observation of Eros yields this quantity with three times the exactness of the earlier methods, and that nothing else than the observation of the lunar orbit for the determination of the parallactic equation can compete with it. Thus the discovery of Eros performed a useful service for astronomy.

Two months after the discovery of Eros, Wolf, still at

Heidelberg, discovered the planet Hungaria (No. 434). This, next to Eros, is, with its solar distance of 1.94 units, the innermost planet of the stream. Its orbit lies, however, entirely in the zone beyond the orbit of Mars.

Just as Eros goes beyond the zone between the orbits of Mars and Jupiter toward the inside, so three lately-discovered planets do so toward the outside; tho, to be sure, to only a small degree. These have also received masculine names.

Highly theoretical investigations have been carried out by Hansen, Glydén, Poincaré and Sir George Darwin, which have gained from orbital computation continually new points of view and have led to beautiful mathematical ideas. Here belongs the question of the simple relations of the times of revolution of a planet and of Jupiter as the most important disturbing planet, or the so-called libration of the orbit, as it comes to light in some measure with Hecuba (No. 108) and with planets of the Hecuba type. The value of the little planets is thus not to be sought upon the practical but rather upon the theoretical side.

But the trouble and time which hundreds of planets uninterruptedly required for their rediscovery and the correction of their orbits was so considerable that the question arose whether the gain was worth all the trouble. So, in the eighties of the nineteenth century, a number of astronomers demanded that a choice be made of the planets, and that only a part be followed up with accuracy, and that the rest (since indeed they apparently showed merely the same phenomena) be abandoned without attention. This would hardly have lessened the labor. For, since the discovery of planets could not be prohibited, the majority of the planets would have been lost and would be continually discovered over again. Professor Tietjen at Berlin found a middle course. Since 1891 he abandoned the yearly ephemerides (tables which give the course of the planets during the entire year) and stated merely in one line, for each planet, time and place of opposition and the daily variation in its position. For selected, important and interesting little planets opposition ephemerides were printed, which gave the position of the planets daily with exactness for six weeks of their best visibility.

The majority of the planetoids appear as mere points and are immeasurably small. However, Weiss estimates, on the basis of brilliancy, the largest of them at 342 kilometers (212 miles) and the smallest at 16.14 and 10 kilometers (10 and 6.2 miles) in diameter.

There is no doubt that many small bodies are in motion among them whose diameters are of 1 kilometer (3,281 feet), of perhaps 1 meter (39.37 inches) and even of 1 millimeter (0.039 inch), and that the majority on account of their faintness are not visible to Earth. The planetoids are of no immediate use to mankind. Their orbital calculations can be used neither for the determination of time nor for the guidance of mariners. But the ideal beauty of these planets is that their reappearance is continually a test of the correctness of mathematical assumptions and that they exhibit over and over again the validity of Newton's law of the universal attraction of all bodies.

CHAPTER XX

THE term "comet," derived from the Latin, coma, or hair, applied to celestial bodies which appear to have a hairy appendage, goes back to the time of the Romans. A similar word, cometa, or cometes, was used by Cicero, Tibullus and other ancient writers. The modern astronomer speaks of the coma of a comet as distinct from the tail, and applies the term to the misty hazy light surrounding on every side a small central bright spot, which he calls the nucleus of the comet.

While the ancients distinguished between comets and meteors, or shooting stars, yet they believed them to be of the same nature, and to be found in the Earth's atmosphere not far above the clouds, or at all events much lower than the Moon. The earlier and Pythagorean view, however, was much more correct, according to modern doctrine; for it held that comets were bodies with long periods of revolution, which idea, like others attributed to Pythagoras, probably came from eastern philosophers of unknown nationality. Apollonius, the Myndian, believed that the Chaldeans were responsible for this notion of the comets; for they spoke of them as travelers that penetrated far into the upper or most distant celestial space. Similar views, typical of the imitative faculty of the Romans, says Humboldt, were held by Seneca and Pliny. The Greek philosophers in many cases preferred to disregard observations. Hence the fanciful theories of Aristotle as regards comets, as well as other astronomical matters, were prevalent for many centuries. Aristotle even believed that the Milky Way was a vast comet which continually reproduced itself.

The comet, with its brilliant head, flaming tail and uncertain appearance, could not be regarded otherwise than as a divine omen to announce some remarkable event or to forebode evil, particularly pestilence and war. Indeed, for many years the death of monarchs was believed, especially by those to whom the wish was father of the deed, to be announced by these brilliant messengers in the sky. In some cases comets were associated with misfortunes, not merely as anticipating or announcing them, but as the actual causes. Seneca's statement that "This comet was anxiously observed by every one, because of some great catastrophe which it produced as soon as it appeared, the submersion of Bura and Helicè" referred to a very brilliant comet which appeared in 371 B.C. about the same time that these two towns of Achaia were swallowed up by the sea during an earthquake. A comet which appeared in 43 B.C. was generally believed to be the soul of Cæsar on its way to heaven. Josephus informs us that the destruction of Jerusalem was announced by several prodigies in 69 A.D., among them a sword-shaped comet which is said to have hovered over the city for the space of a year! Another classical instance may be quoted from Pliny (23-79 A.D.) in his "Natural History" where he says: "A comet is ordinarily a very fearful star; it announces no small effusion of blood. We have seen an example of this during the civil commotion of Octavius." When the comet of 79 A.D. appeared, the Roman Emperor Vespasian refused to be intimidated by the frightening interpretation placed upon it. "This hairy star does not concern me," he is reported to have said; "it menaces rather the King of the Parthians, for he is hairy and I am bald." Not long after the appearance of the comet he died. No doubt the prophecies of the imperial sooth-sayer were more highly regarded thereafter.

The ancient Greeks and Romans were not the only ones who took these heavenly apparitions seriously. In France the great eclipse of 840 was said to have hastened the end of Louis le Debonaire, and it was firmly believed that the comet which appeared a year or two previously presaged this occurrence. Much of the mysticism attaching to figures found expression in the superstition that the Christian era could not possibly run into four figures. Hence the end of the world was looked for by many of the inhabitants of Europe when the year 1000 approached. So widespread was this belief that husbandry and toil were neglected. When a comet appeared the feeling was strengthened. Nothing remarkable occurred, however, beyond the natural consequences of such wholesale neglect of the proper care of the soil. Famine and pestilence in succeeding years were the result.

The comet, which, we shall see, was the famous comet of Halley that blazed forth in April, 1066, was believed to presage the success of the Norman Conquest, and the invasion of England by the Normans "guided by a comet" was a familiar topic for the chroniclers of the time. The abdication of Emperor Charles V was reported to have been influenced by the comet of 1556, but the event had already taken place before the comet made its appearance. Gian Galeazzo, the Visconti Duke of Milan, viewed the comet of 1402 as a celestial sign of his approching death.

A striking example of the manner in which the comet was regarded is contained in the contemporaneous description by Ambroise Paré, the father of French surgery (1517-1590), in which he speaks of the fear inspired by the comet of 1528. "This comet," said he, "was so horrible, so frightful, and it produced such great terror in the vulgar, that some died of fear and others fell sick. It appeared to be of excessive length, and was of the color of blood. At the summit of it was seen the figure of a bent arm, holding in its hand a great sword, as if about to strike. At the end of the point there were three stars. On both sides of the rays of this comet were seen a great number of axes, knives, blood-colored swords, among which were a great number of hideous human faces, with beards and bristling hair."

The comet of 1472, apparently, was the first comet to receive scientific study and not be regarded solely as a cause of superstitious terror. A series of observations were made in Franconia by Johann Müller, of Königsberg, known as Regiomontanus.

By the time of Tycho Brahe, while comets were not satisfactorily explained, yet they were being considered on a more rational basis, and the correctness of the Aristotelian doctrine was, as in other matters, being questioned. It was believed before this that comets were generated in the higher regions of the atmosphere. But in 1507, on the appearance of a brilliant comet, Tycho, in an elaborate series of observations, satisfied himself that the strange body was at least three times as far off as the Moon and also that it was revolving around the Sun in a circular orbit at a distance greater than that of Venus. Comets subsequently were observed by Tycho and his pupils. His observations in this field led to those ideas of the solar system of his which we have discussed. It was but natural that Kepler, as a follower of Tycho, should have paid especial attention to comets. In 1607 he observed the comet now known as Halley's comet. Kepler believed that comets were celestial bodies which move in straight lines and after having passed the Earth recede indefinitely into space. Assuming that these strangers in the heavens would never reappear, he did not consider that their paths required serious study, for which reason he made no observations to ascertain their movements and test his theory. Before Kepler, Jerome Fracastor (1483-1543) and Peter Apian (1495-1552) had observed that a comet's tail always points away from the Sun, no matter in what direction it may be traveling, and with this observation Kepler agreed, adding as an explanation the supposition that the tail was formed by rays of the Sun penetrating the body of the comet and carrying away with them some portion of its substance. This theory, after due allowance has been made for the change in our conception of the nature of light, is of interest as an anticipation of the modern theory of comets' tails. Kepler found himself compelled in his "Treatise on Comets," 1619, in which the foregoing observations were published, to refer to the meaning of the appearance of a comet and its influence on human affairs. At this time there were striking events enough in the affairs of Europe to prove any theory of the influence of comets on human life. He realized, however, that comets are very numerous, for he states, "There are as many arguments to prove the motion of the Earth around the Sun as there are comets in the heavens."

The motion of the comets was also studied by Galileo in 1623 as a part of the motion of the Earth in the Copernican theory of the solar system. But the first and most important contribution to the true explanation came from Dörfel of Saxony, who proved from the comet of 1681 that the orbits of comets are either very elongated ovals or parabolas and that the Sun occupied a focus of the curve. Newton, discussing this subject in his Principia, reached independently the same conclusion a few years later and established it as a universal law by incontrovertible mathematical proof.

By the seventeenth century a considerable number of comets had been recorded. John Hevel, of Danzig (1611-1687), published two large books on comets, Prodromus Cometicus (1654) and Cometographia (1668), which contained the first systematic account of all recorded comets.

It was a brilliant thought of Newton's that led him to consider whether gravitation toward the Sun could not explain a comet's motion just as well as that of the planet, and if so, as he took pains to prove in the beginning of the Principia, such a body must move either along an ellipse or in one of two other allied curves, the parabola and hyperbola.

Edmund Halley (1656-1742), who had been a friend and active associate of Newton's and had assisted him for several years in the preparation of the Principia, followed

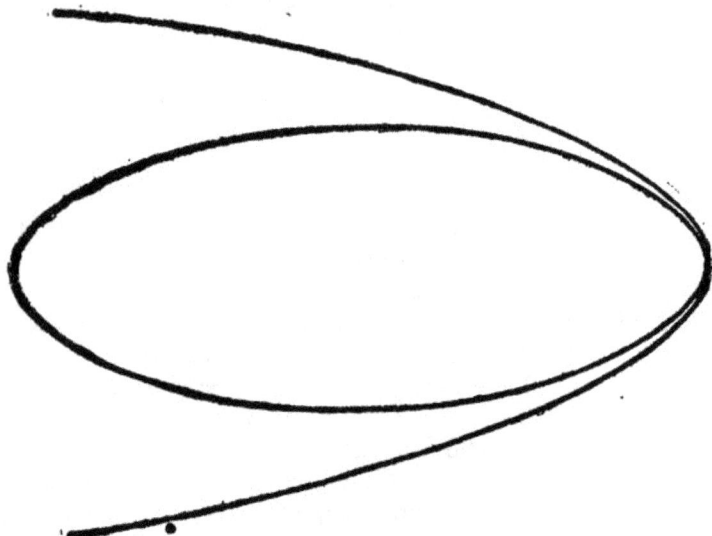

Fig. 29 —An Elongated Ellipse and a Parabola.

Newton's principles in the observation of comets. He computed the paths of the comets of 1680 and 1682, and especially one of 1531 whose appearance was recorded by Appian. His studies contributed much to the material dealing with this subject in the Principia, especially the later editions. In 1705 he published a Synopsis of Cometary Astronomy, in which he calculated 24 cometary orbits. Discussing in detail a number of these, he was struck with the resemblance between the paths described by the comets of 1531, 1607 and 1682, and the approximate equality in the intervals between their respective appearances and

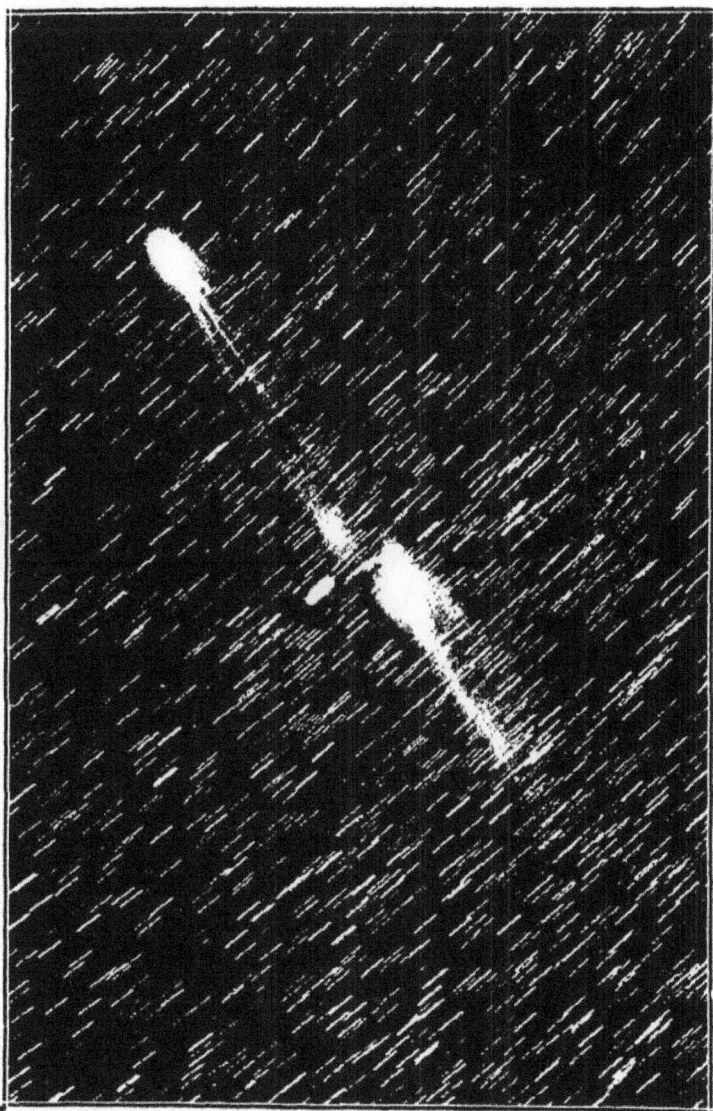

MOREHOUSE COMET, OCTOBER 1, 1908.
The stars are shown as streaks owing to camera having followed
comet's path.   (Yerkes Observatory.)

that of a fourth comet observed in 1456. Moreover, there was historical record of a comet in 1380, as well as in 1305. He at once concluded that all four comets were really different appearances of the same comet, which moved around the Sun in an elongated ellipse in a period of about 75 or 76 years, and he accounted for the small differences in the different intervals between the appearances of the comet by perturbations caused by planets in whose neighborhood the comet passed. He then made the

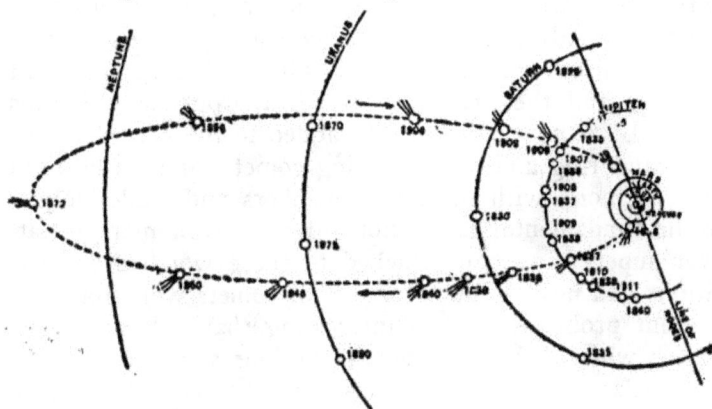

Fig. 30 —HALLEY'S COMET DURING CYCLE 1835-1910.
Note: The places given are for January 1 of the dates indicated.

first prediction of the probable reappearance of the comet, assigning the date, 1758, when next it would be seen after a 76-year interval. Halley, no less than modern astronomers, was aware of the disturbance that the presence of the planets might work in a comet and its orbit, and how its time might be altered. He confidently announced the actual appearance of the comet, but left shortly before his death, at the age of 85 years, the following quaintly worded statement in regard to the comet: "Wherefore, if according to what we have already said, it should return again about the year 1758, candid posterity will not refuse to acknowledge that this was first discovered by an Englishman." When the time arrived the comet was looked for by astronomers; the French savant, Alexis Claude Clairaut (1713-1765), computed the various perturbations which might have affected its journey. As its path lay through the orbits of Jupiter and Saturn and as it passed close to both of these great planets, his calculations showed that there might be expected a retardation of 100 days on Saturn's account and 518 days for Jupiter. On Christmas Day, 1758, a month and a day before the date assigned by Clairaut, and in the year announced a half century before by Halley, the comet was actually discovered by George Palitzsch (1723-1788), of Saxony, and the great astronomical prophecy was thus fulfilled. A new member was added to the solar system; the wandering and fear-inspiring comet was thus brought into harmony with the other members and made subject to the fundamental calculations of the astronomer. Whatever superstition had attached to these wonderful apparitions had now all but passed, and comets were found to present problems no less interesting than other celestial bodies when their fundamental motions were known.

In 1835 Halley's comet duly reappeared and passed through its perihelion within a few days of the time set for it by astronomers. It was observed, among others, by Sir John Herschel at the Cape of Good Hope. In 1910 this comet returns again along the orbit shown in Fig. 30. It was first discovered by Wolf, of Heidelberg, on September 11, 1909.

In the study of Halley's comet in connection with its appearance, much attention has been devoted by astronomers to its earlier history, particularly that recorded in Chinese and European annals. Messrs. Cowell and Crommelin, at the Greenwich Observatory, have carefully examined the previous work of Hind and have found it in the main correct. Halley's comet unquestionably must be identified with one that occurred in 1066, in the year of the Norman Conquest, a representation of which is now extant in the Bayeux tapestry supposed to have been worked by Queen Matilda and her ladies.

In more modern times comets have been associated with some important development of scientific theory rather than with historical events. The comet of 1811, visible from March 26th of that year until August 17th of the following year, received the attention of Sir William

Fig. 31 —THE COMET OF 1066 AS REPRESENTED IN THE BAYEUX TAPESTRY. (From the World of Comets.)

Herschel, who discovered that it shone partly by its own light, which developed as it approached the Sun. This comet had a tail at one time 100 million miles in length and 15 million miles in diameter. Dr. Olbers suggested that electrical repulsion was responsible for the formation of the tail. A comet famous for the fact that it was the first of the family of Jupiter's comets to be discovered was that named for Johann Franz Encke, for many years

director of the Berlin Observatory. It was discovered by
Pons of Marseilles, November 26, 1818, but in the calculation of its orbit and other elements Encke found that
it revolved about the Sun in a period of 3⅓ years, which
is considerably shorter than that of any other known

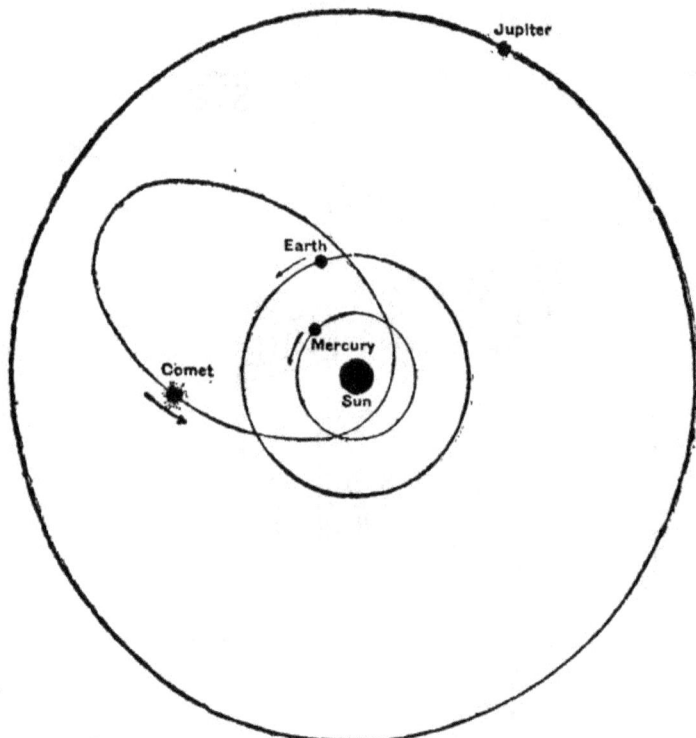

Fig. 32 — THE ORBIT OF ENCKE'S COMET.

comet. Furthermore, he established its identity with
comets seen by Méchain in 1786, by Caroline Herschel in
1795, and by Pons, Huth and Bouvard in 1805. Encke's
calculations, after establishing its periodicity, assigned the
date of May 24, 1822, for its next return to perihelion,
and tho on account of the position of the Earth at

that time it was invisible in the northern hemisphere, it was detected at Sir Thomas Brisbane's observatory at Paramátta by Rümker very nearly in the position indicated by Encke. This was only the second instance of the recognised return of a comet, so that Encke's work as an astronomical achievement should be considered with that of Halley.

Biela's comet, discovered by an Austrian officer of that name at Josephstadt, in Bohemia, February 27, 1826, presents as interesting features as that of Encke. It was seen ten days later at Marseilles by the French astronomer Gambart. Both observers announced its discovery and the computation of its orbit in the same issue of the Astronomische Nachrichten. Tho the comet was identified with similar appearances in 1772 and 1805, it was not visible after the latter date with the naked eye. In 1832 Sir John Herschel observed it as a conspicuous nebula without a tail. While the day of active superstition in regard to the appearance of comets had passed with the demonstration of Halley's theory, nevertheless Biela's comet occasioned widespread popular excitement, founded, moreover, on the statements of scientific men that a collision with the Earth might occur. The possibility is one of the remotest of cosmical happenings.

In 1846 Biela's comet reappeared, and when first seen, November 28, presented no unusual appearance. Gradually it became distorted, however, and elongated. Within two months it divided into two separate bodies, which were visible until April 16th of the following year. This striking phenomenon of a double comet was noted by many astronomers at different observatories, and thus established what Seneca had reproved Ephorus for supposing to have taken place in 373 B.C. and what Kepler had noted in 1618, but without convincing astronomers at large of the correctness of his impression. These two Biela comets contained a small amount of matter and performed their revolutions around the Sun independently without experiencing any appreciable mutual disturbance, which indicated that at an interval of only 157,250 miles their attractive power was virtually inoperative. Various interesting phenomena showing internal agitation were observed and variations of brilliancy and form were distinctly evident. In 1852 Biela's comet again appeared in its double form, but since that time has not been observed. The disruption occasioned by its proximity to Jupiter in 1841 is believed to have been the beginning of the disintegrating process which resulted in its disappearance.

The greatest comet of the nineteenth century was that of Donati, which was seen by him at Florence, June 2, 1858. By the end of September, when the comet had reached its perihelion, the tail had attained full development, and on October 10 it stretched in a maximum curve over more than a third of the visible hemisphere, representing a length of 54,000,000 miles. For the 112 days during which it was visible to the naked eye the fullest observations were made. The comet stands by itself, as it is not possible to identify it with any other. At aphelion its orbit extended out into space to 5½ times the distance from the Sun to Neptune, and for its circuit, which is effected in a retrograde direction, requires more than 2,000 years, so that its next return should be about the year 4000 A.D.. It was computed by M. Faye that the volume of this comet was about 500 times that of the Sun. On the other hand, he calculated that the quantity of matter it contained was only a fraction of the Earth's mass. This shows how almost inconceivably tenuous the material forming the comet must have been—much more rarefied indeed than the most perfect vacuum which can be produced in an air pump. This tenuity is shown by the fact that stars were seen through the tail "as if the tail did not exist." A mist of a few hundred yards in thickness is sufficient to hide the stars from our view, while a thickness of thousands of miles of cometary matter does not suffice even to dim their brilliancy.

Dr. Barnard states that our knowledge of the extremely rapid transformations in the tails of comets dates from the photographs of Swift's comet of 1892, taken at the Lick Observatory, and similar ones taken of the same object by Professor Pickering at Arequipa. While only an insignificant affair visually, and but fairly visible to the naked eye, Swift's comet showed upon the photographic plates the most extraordinary and rapid transformations. One day its tail would be separated into at least a dozen individual streams and the next present only two broad streamers, which a day later had again separated into numerous strands, with a great many apparently a secondary comet, appearing some distance back of the head in the main tail, with a system of tails of its own.

The photographs of Brook's comet of 1893 showed such an extraordinary condition of change and distortion in the tail as to suggest some outside influence, such as the probable collision of the tail with a resisting medium, possibly a stream of meteors. The long series of photographs obtained of this comet showed great masses of cometary matter drifting away into space, probably to become meteor swarms. Had it not been for photography, the comet, instead of proving to be one of the most remarkable on record, would have passed without special notice. Tho these phenomena were so conspicuously shown, scarcely any trace of the disturbance was visible with the telescope. On account of the apparent insignificance of the comet visually, no photographs were made of it elsewhere during its active period. The application of photography to cometary studies has been an important feature of the investigation of later comets, none of which since 1882 have been marked by great brilliancy.

The general nature and appearance of a comet is thus clearly described by the late Professor R. A. Proctor: "When first seen in a telescope, a comet usually presents a small round disk of hazy light, somewhat brighter near the center. As the comet approaches the Sun the disk lengthens, and if the comet is to be a tailed one, traces begin to be seen of a streakiness in the comet's light. Gradually a tail is formed, which is turned always from the Sun. The tail grows brighter and longer, and the head becomes developed into a coma surrounding a distinctly marked nucleus. Presently the comet is lost to view through its near approach to the Sun. But after a while it is again seen, sometimes wonderfully changed in aspect through the effects of solar heat. Some comets are brighter and more striking after passing their point of nearest approach to the Sun (or perihelion) than before; others are quite shorn of their splendor when they reappear. On the other hand, the comet of 1861 burst upon us in its full splendor after perihelion passage.

"As a comet approaches the Sun a change takes place in the appearance of the coma and nucleus, and in some instances a tail is generated. The process actually observed is generally this: In the forward part of the nucleus a turbulent action is seen to be in progress, leading to the propulsion toward the Sun of jets or streams of misty-looking matter. Sometimes a regular cap or envelope is seen to be projected in this manner toward the Sun, or even a set of envelopes one within the other. The matter thus thrown off is not suffered to pass very far from the nucleus toward the Sun, but is swept away, as fast as formed, in the contrary direction. If the funnel of a steam engine were directed forward, instead of upward, then the appearance presented by the emitted steam as the engine rushed on (against a hurricane, suppose, to make the illustration more perfect) would exemplify the process which seems to be taking place around the front of the nucleus and far behind it as the matter formed is continually swept away from the Sun. The same Sun which attracts the nucleus seems to repulse the emitted matter with inconceivable energy.

"When we see the tail of a comet occupying a volume

thousands of times greater than that of the Sun itself, the question naturally suggests itself: 'How does it happen that so vast a body can sweep through the solar system without deranging the motion of every planet?' Conceding even an extreme tenuity to the substance composing so vast a volume, one would still expect its mass to be tremendous. For instance, if we supposed the whole mass of the tail of the comet of 1843 to consist of hydrogen gas (the lightest substance known to us), yet even then the mass of the tail would have largely exceeded

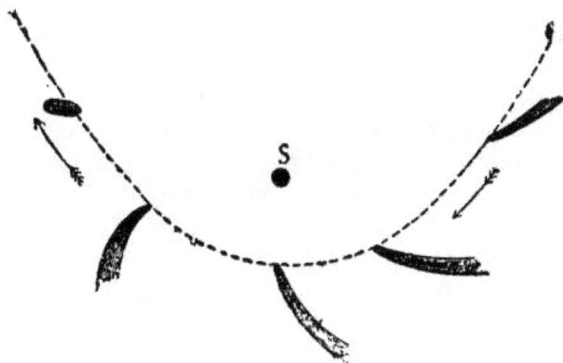

Fig. 33 —The Tail of a Comet Directed from the Sun.

that of the Sun. Every planet would have been dragged from its orbit by so vast a mass passing so near. We know, on the contrary, that no such effects were produced. The length of our year did not change by a single second, showing that our Earth had been neither hastened nor retarded in its steady motion round the Sun. Thus we are forced to admit that the actual substance of the comet was inconceivably rare. A jarful of air would probably have outweighed hundreds of cubic miles of that vast appendage which blazed across our skies to the terror of the ignorant and superstitious.

"The dread of the possible evils which might accrue if

the Earth encountered a comet will possibly be diminished by the consideration of the extreme tenuity of these objects. But the feeling may still remain that influences other than those due to mere weight or mass might be exerted upon terrestrial races in the course of such an encounter. On account of their enormous volumes, it is not so utterly improbable that we should encounter them as that we should meet the comparatively minute nuclei. In fact, the Earth actually did pass through the tail of the comet of 1861. At about the hour when it was calculated that the encounter should have taken place a strange auroral glare was seen in the atmosphere, but beyond this no effect was perceptible."

In distinction to the comets moving in regular orbits around the Sun, the possible portions of one much larger cometary body which became dispersed by gravitational action or through violent encounter with the suns surrounding must be mentioned. These comets, which apparently have been seized by the gravitative attraction of planets, are compelled to revolve in short ellipses around the Sun well within the limits of the solar system. These comets are spoken of as "captures," and while Jupiter, Saturn, Uranus and Neptune each possess families of this kind, it is the first named which are the most important, as they number about 30, and in this family may be included not only the bodies that Jupiter has attracted, but those that have been robbed from other planets. Comet families are not found in the case of the terrestrial planets, because the gravitative power of the Sun in their vicinity is so much greater than any attractive force which they could manifest. In addition, when a comet enters the inner portion of the solar system, it has such velocity that the gravitational attraction of the planets within these regions is not powerful enough to cause any appreciable deflection. If a captured comet is acted upon by further disturbing causes, its new orbit may be disarranged and it may be again diverted into celestial space. The nature of these orbits is shown in Fig. 34.

The facts learned from modern cometary study are summed up by Miss Clerke as follows: "First, comets may be met with pursuing each other, after intervals of many years, in the same, or nearly the same, track; so

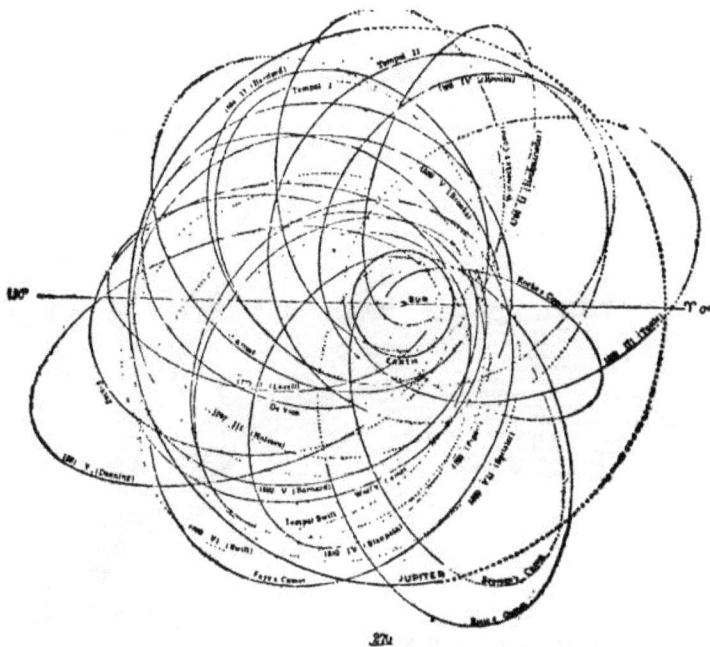

Fig. 34 —JUPITER'S FAMILY OF COMETS.

The dotted portions represent the parts of the orbits below, s. e., to the south of the ecliptic.

that identity of orbit can no longer be regarded as a sure test of individual identity. Secondly, at least the outer corona may be traversed by such bodies with perfect apparent immunity. Finally, their chemical constitution is highly complex, and they posssess, in some cases at least, a metallic core resembling the meteoric masses which occasionally reach the Earth from planetary space."

The first serious study of the physical nature of comets' tails was undertaken in 1811 by Dr. Heinrich Olbers, the astronomer of Bremen. He assumed that the formation of a tail was due to expelled vapors on which two forces, solar and cometary, acted and balanced each other. In

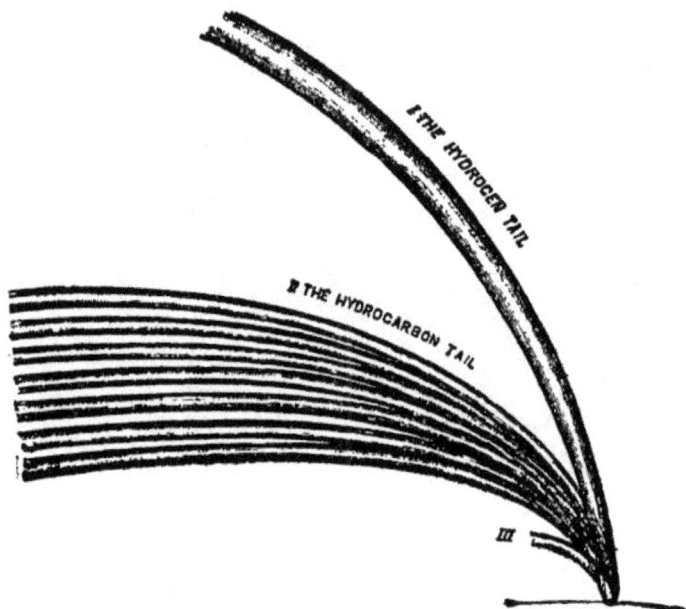

Fig. 35 —BREDICHIN'S THEORY OF COMETS' TAILS.

other words, he believed that the tails were emanations, not appendages, and consisted of rapid outflows of highly rarefied matter which, in great part, had become permanently detached from the nucleus.

This theory is especially interesting in the light of modern investigation. It served for many years until that of Bredichin, who in an examination of various comets' tails found that the curvilinear shapes of the outline fall into one or an other of three special types as indicated in the accompanying illustration. Type 1, or the straightest, is most probably due to the element hydrogen. In Type 2 a number of hydrocarbons are present in the body of the comet, while in the third type iron or some element with high atomic weight was assumed. Some comets may have tails of more than one of these forms, as, for example, in the case of Donati's comet, which had a straight as well as a curved tail. Of Type 2, comets with a number of tails have been recorded, such as the one of the year 1744. Bredichin calculated that a repulsive force adequate to produce the straight tail of Type 1 need only be about 19 times as much as the attraction of gravitation, while tails of the second type a repulsive force about equal to 3.2 to 1.5 times that gravitation would suffice, while those of the third type would require a repulsive force about 1.3 to 1 times that of gravitation. These parts are nearly inversely proportional to the atomic weights of hydrogen, hydrocarbon gas and iron vapor, which ratio suggested to Bredichin the composition of the various types of tail. While he was unable to demonstrate that the tails were the result of electrical action, yet he assumed some hypothetical repulsive force which electrical action seemed to explain better than any other.

Professor E. E. Barnard in 1905 stated that a repellent influence of some sort must come from the Sun, and with it he included an ejecting force proceeding from the comet itself and a resistant force of some kind. The repellent force from the Sun may be found in the pressure of light, which Professor J. Clerk Maxwell assumed must be exerted by light rays according to mathematical reasoning. "Radiation pressure," as it is termed, was not experimentally proven for many years, but in 1900-1901 it was established as a scientific fact by the Russian physicist Lebedev and in America by Nichols and Hull. This principle thus demonstrated, Professor Svante Arrhenius

applied cosmically and held responsible for the generation of streams of matter flowing from the comet's head.
As the comet approaches the Sun this pressure exceeds
the force of gravity and acts upon the cometary substance
so as to drive out multitudes of the minute particles in a
direction away from the Sun. Such a swarm of particles
receiving the light from the Sun would appear as the
familiar luminous streamer recognised in the comets' tails.

When examined with the spectroscope, a comet's tail
shows a faint continuous spectrum, produced doubtless by
the sunlight reflected by the small particles, in addition to
spectral bands due to gaseous hydrocarbons and cyanogen.
Cyanogen is due to electric discharges, for such discharges
are observed in comets whose distance from the Sun is so
great that they cannot appear luminous owing to their own
high temperatures. In other words, the composition of a
comet is not unlike the blue flame of a gas stove, which is a
combination of hydrogen and carbon. As the comet
dashes toward the Sun and its temperature consequently
rises, the spectroscope reveals the presence of iron, magnesium, and other metals in the nucleus. With a closer
approach to the Sun the hydrocarbons split up into hydrogen gas and hydrocarbons of a higher boiling-point.
Finally, a time comes when these more refractory hydrocarbons in turn decompose into free carbon in the form
of soot. Because interstellar space is airless the soot
cannot burn, but must accompany the comet in the form
of a very fine dust. This dust, propelled away from
the Sun by radiation pressure, constitutes the tail of many
a comet. Some of the soot particles may be larger
than the critical size. They will be jerked forward toward
the Sun in advance of the comet to form what is known
as the comet's "beard," a rather rare phenomenon.

This phenomenon of the pressure of light is able to
explain the fact that the minute particles ejected from
the nucleus of a comet can pass over great distances in
small intervals of time, which was one of the hardest

points to overcome in explaining the rapid change of position of the comet's tail passing around the Sun.

In addition to the light-pressure of the Sun, the electrical energy of the Sun must be called upon to explain the occurrence of tails which are ejected from the nucleus with a force that may be as much as 40 times more powerful than gravitation.

Meteors, often called shooting stars, which is some instances, at least, must be the remains of comets, are small solid bodies which revolve around the Sun, generally in great numbers, following approximately the same orbit, and are encountered by the Earth in its annual revolution. Then they graze the Earth or even fall toward it, but, fortunately for its inhabitants, they seldom reach its solid surface, because they are raised to incandescence and dissipated in vapor by the heat generated by friction in their swift rush through the atmosphere. At certain seasons of the year the Earth traverses comparatively dense swarms of meteors and is subjected to a veritable bombardment. The effect to the eye of these flashing meteors is most striking and brilliant, particularly if the point of the swarm intersected contains a large aggregation of meteors. The apparent radiation of a meteoric shower from a common point or radiant is an effect of perspective, as the meteors of a swarm in reality pursue parallel paths. Three of these meteor swarms are of particular interest, as under certain conditions they give rise to fine displays. These are known as the Bielids, the Leonids, and the Perseids, the first two occurring in November and the last named in August.

The Bielids, deriving their name from the connection of their orbit with that of the comet Biela, are also known as Andromedids on account of their apparent source in the constellation of Andromeda. M. Egenitis, Director of the Observatory at Athens, traced back the Andromedid shower to the times of the Emperor Justinian. Theophanes, the chronicler of that epoch, writing of the famous revolt of Nika in the year 532 A.D., says: "During the same year a great fall of stars came from the evening till the dawn." M. Egenitis notes another early reference to these meteors in 752 A.D., during the reign of the eastern Emperor Constantine Copronymous. Writing of that year, Nicephorus, a Patriarch of Constantinople, states: "All the stars appeared to be detached from the sky and to fall upon the Earth." But

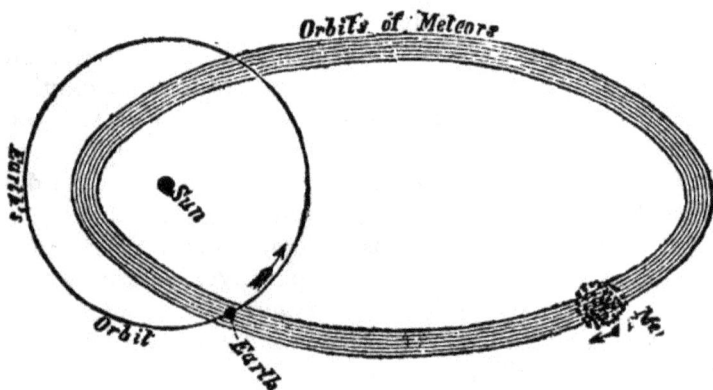

Fig. 36 —THE ORBIT OF A SHOAL OF METEORS.

it was not until the nineteenth century that Bielids aroused much attention, and then it was in great part due to the fact that apparently the same orbit was occupied by them as by Biela's comet, which we have seen was not observed after its appearance in 1852. The Bielid shower, however, since that time has shown increased activity, which was especially true in the years in which the comet, were it in existence, would have been scheduled to pass near the Earth.

In the case of the Leonids, records of their occurrence go back as far as 902 A.D., which is called "year of stars," because on the night of October 12, while the Moorish Ibrahim Ben Ahmed was dying before Cosenza in Calabria, "a multitude of falling stars scattered themselves across the sky like rain," and naturally aroused great excitement among those who beheld the phenomenon, which they considered a celestial portent of unusual significance. In 1698 modern history of the Leonids began. A maximum Leonid shower has occurred with considerable regularity at periods of about 33 years from that date. In 1799, on

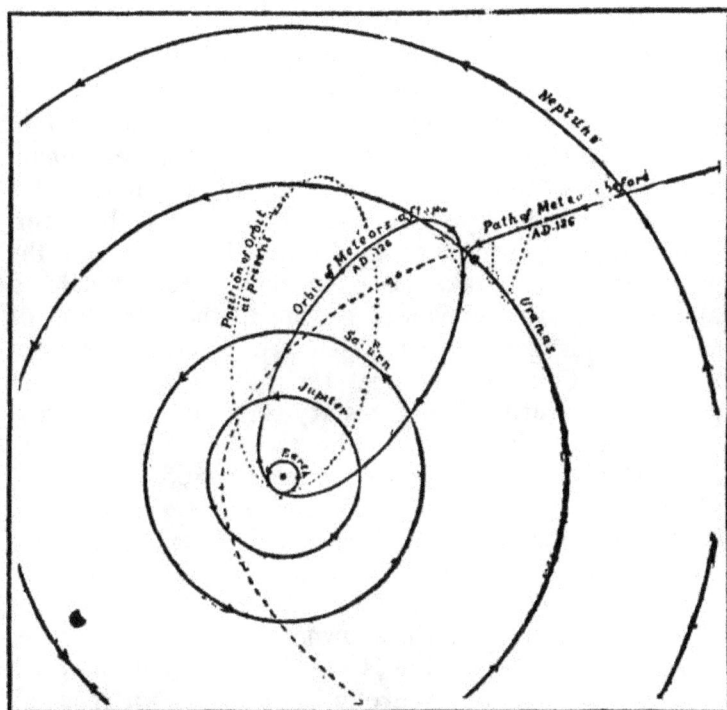

Fig. 37 —THE HISTORY OF THE LEONIDS.

the 11th of November, Humboldt and Bonplandt witnessed a notable display in South America. On November 12, 1833, meteors were said to have fallen as thickly as snow-flakes; in seven hours 240,000 were estimated to have appeared. The radiant from which the meteors seemed to come was found to be situated in the head of the constellation of Leo, from which circumstance the name Leonids results. Professor Dennison Olmstead, of Yale University, assigned to this cloud of cosmical particles the path of a narrow ellipse in an orbit around the Sun and intersecting that of the Earth. This marked the beginning of an important department of astronomy. In 1837 Olbers established the periodicity of the maximum shower which indicated a regular distribution of the meteoric supply. In 1866, as conjectured by Olbers, another time of maximum occurrence took place, which seemed to demonstrate that while the Earth cut through the orbit each year about the same date, at the 33-year period the swarm was at a point of maximum density in the orbit. In 1899, however, much disappointment was caused by the failure of the Leonid shower to take place on the scheduled date, a failure which was explained as due to the attraction of one of the larger planets which had diverted the orbit from its old position so that the Earth failed to pass through the swarm. The cometary connection in the case of the Leonids is shown by the fact that they seem to travel in the orbit of Tempel's comet of 1866.

The Perseids date back to the year 811 A.D., and derive their name from the constellatio of Perseus, where their radiant point is situated. They are seen on the 10th of August in continental Europe. As this is the day of Saint Lawrence, they are known as the "tears of Saint Lawrence." But this date is not the only one on which meteors from this swarm are to be observed, for they fall in greater or less numbers from about July 8th to August 22d. They are very rapid in their motion and the trails often persist for a minute or two before they are disseminated. The Perseids have an easterly motion, shifting each night by a small amount. Their orbit cuts the orbit of the Earth almost perpendicularly, and they are supposed to be the débris of an ancient comet which traveled the same path. Various comets, especially that of Tuttle in

1862, seem to have had the same orbit, and the meteors are quite evenly distributed along the path.

There are other swarms of meteors, such as the Lyrids, the Orionids, etc., and Mr W. F. Denning, of England, who has made a specialty of this field, has accounted for 3,000 other less conspicuous showers. Astronomers have shown that the various meteor swarms and comets move in the same orbits. Accordingly the theory has been proposed that when a comet is captured by a planet the material of the tail is driven off into space and the remaining material, disintegrated by the various forces at work, is distributed along the orbit. Consequently the phenomenon of a meteoric shower occurs when the orbit of a swarm and that of the Earth intersect and when the Earth and the meteors arrive at this intersection at the same time. Leverrier showed that the Leonids resulted from the capture of a parent comet in 126 A.D. at the time of a near approach, and that the disintegration, not entirely completed, is already far advanced. He claimed that the Perseids were of much older formation.

Meteors have been observed from the earliest days, but are of such minor importance as compared with comets that they attracted no particular attention. In 1719 Brandes and Benzenberg, at Göttingen, by making simultaneous observations of the beginning and end of the path of a meteor from different stations a few miles apart were able to determine not only its position, but its velocity, and subsequent observations similarly made indicate that meteors appear at altitudes of 60 to 100 miles and that they move over paths of 40 or 50 miles, traveling at a rate of 10 to 40 miles a second.

When a meteor enters the Earth's atmosphere from interplanetary space, the friction of the atmosphere, caused by its high velocity, develops heat and causes it to shed a brilliant light. The temperature of a meteor rises to many thousand degrees Centigrade, and for that reason it is usually consumed before it reaches the Earth's surface. The products of oxidation and disintegration consist simply of dust, which falls on the Earth's surface or is distributed throughout space. In the main, meteors simply contribute dust to the Earth.

The energy of the meteor as well as its mass can readily be ascertained if its distance, the duration of its luminosity, and its brightness are known, for the total amount of light radiated can be calculated on the basis that its entire energy is thus transformed. The mass of a meteor is not particularly large and is usually but a small fraction of an ounce. For the most part it is not larger than a pea or pebble. It is the atmosphere that not only heats these rapidly moving bodies but acts as a protection to the Earth, for if meteors were not thus disintegrated they would fall upon its surface in a constant bombardment. In addition to the meteors seen with the naked eye, estimated by the late Professor Simon Newcomb at not less than 146 billion per annum, there are doubtless ten times as many which pass merely as streaks of light in the field of the observer's telescope.

In addition to the extremely fine dust which settles on the Earth as the result of the disintegration of various celestial bodies, there are from time to time masses of greater or less size which, rushing into the Earth's atmosphere with a brilliant glow due to the heat generated by friction, fall to the Earth's surface and become more or less embedded. The appearance is most striking, accompanied as it often is by a loud roar like a waterfall and occasionally violent explosions. Thus it is stated that at Cairo, in August, 1029, many stars passed with a great noise and a brilliant light. These bodies, a number of which come to the Earth's surface yearly, are termed meteorites, siderites, uranoliths or aerolites, and apparently are the connecting links between the Earth and outside space. Their nature is none too well known and they present many unsolved problems. It is interesting to know that the great Mexican meteorites at the time of the Spanish invasion were considered holy bodies by the Indians, so that it is inferred that their fall from the heavens was known and was regarded as a supernatural occurrance. In the Greek and Roman records similar attention was paid to the palladium of Troy, to the image of Diana at Ephesus and to the sacred shield of Numa, all of which were said to have fallen from the heavens and were no doubt meteorites.

Meteorites are usually divided into two classes, those composed chiefly of iron and those composed chiefly of stone. Of the 292 actually observed meteoric fall that took place during the last century, only twelve, or about 4 per cent., belonged to the first class, yet in our cabinets the two classes are represented in nearly equal numbers. The explanation of this strange anomaly lies in the fact that unless the fall has been actually witnessed close at hand, very few of the stony meteorites are ever found. Of 328 in the collection of the British Museum, 305, or 93 per cent., were seen to fall. This is partly because these bodies to ordinary inspection appear very like common stones, and therefore are not recognised as meteorites, and partly because owing to their physical and chemical structure they are readily decomposed by the action of the elements.

It is the custom to associate meteorites with falling stars, and to say therefore that they are of cometary origin. This relationship, however, is not as obvious, when we begin to examine into the case, as at first sight appears. A prominent difficulty is that the distribution of the meteorites throughout the year differs very materially from that of the falling stars and fireballs. While these last two are about twice as numerous during the latter half of the year as during the first half, the meteorites are more numerous during the first half of the year. From this we should infer that while perhaps all meteorites are fire-balls, only comparatively few fireballs become meteorites.

The dividing line between meteorites and falling stars then lies among the fireballs, the swiftly moving ones being allied to the falling stars and the slowly moving ones to the meteorites.

It is now generally accepted that the crystalline and often conglomerate structure of these bodies proves them to be but the fragments of much larger bodies that have in some manner been destroyed or from which they have otherwise become separated. Many believe that the crystalline structure of the iron meteorites indicates a slow cooling, while some say that the structures of the "chondres" of the stony meteorites must certainly have been produced by a very rapid crystallization due to a sudden exposure to a lower temperature.

It was formerly thought by some that these bodies might have been expelled from the Sun. Altho it is quite possible that solar explosions in past ages were sufficiently violent to project these bodies with the necessary cometary velocities, yet we cannot believe the Sun to be the direct source of them, since it is improbable that either solid stone or iron should ever have existed upon its surface or within its interior. Nor is it easy to explain how with such an origin the meteorites should have acquired their present orbits.

Some of the earlier cosmogonists referred their origin to the terrestrial or lunar volcanoes. This is manifestly impossible in the case of the Earth, since even prehistoric volcanoes could not have expelled their products with such force that after leaving the confines of our atmosphere they should still retain a velocity of over seven miles per second. Yet this is the speed required to prevent an immediate return to the Earth's surface. Moreover, altho volcanic eruptions in prehistoric times were undoubtedly more frequent and voluminous than at present, it is by no means certain for theoretical reasons that they were then any more violent than they are to-day.

Meteors escaping from lunar volcanoes would not have

to encounter a dense atmosphere, and, furthermore, their required parabolic velocity would be appreciably less. But even under the most favorable circumstances, in order to escape both the Moon and Earth a speed of over two miles per second would be required. That attained, they would then be controlled by the Sun and might be picked up at any later time by our planet in its orbit. The objection to this explanation is that no explosive volcanoes have ever been detected upon the Moon, all the craters being of the engulfment type. It is therefore very improbable that such extremely violent explosions could have occurred there.

While, as we have seen, meteorites cannot be the product of terrestrial volcanoes, yet it is suggested by Prof. W. H. Pickering that the stony ones were all of them formed during the great cataclysm that occurred at the time that the Moon separated from the Earth. When the truly enormous pressure on the deep-lying terrestrial strata was suddenly relieved by the departure of the upper layers, which now form our Moon, tremendous explosive energy must have been generated. Considerable portions of our atmosphere must have followed the larger flying masses, and the atmospheric resistance to the smaller ones, swept along at the same time, would have been much diminished. Altho we can probably never definitely know just what occurred at this time, it is quite possible that considerable quantities of the smaller masses were carried along by the blast of escaping gases and were projected to such distances as to free themselves entirely from the attraction of our planet. This implies a solid crust for the Earth at the date of birth of the Moon, which previous investigations of the place of origin of that body seem to justify.

In support of this view of the terrestrial origin of meteorites we have the fact that twenty-nine terrestrial elements, including helium, have so far been recognised in them, ten of them being non-metallic. No new elements have been found. The six which occur most frequently in the Earth's crust, named in the order of their abundance, are oxygen, silicon, aluminium, iron, calcium and magnesium. The eight most commonly found in the stony meteorites are these six, besides nickel and sulphur.

Nearly all the stony meteorites contain some metallic iron, and some of them contain large quantities of it. But this is also true of some of our basaltic lavas. Indeed, large masses of iron have been found in ledges upon the Greenland coast. Some of this iron contains over 6 per cent. of nickel, but much larger proportions have been discovered in New Zealand, Piedmont and Oregon, where considerable quantities of the nickel iron alloys have been found. According to Farrington, of the twenty-one minerals recognised in meteorites, fourteen have been found in our volcanic products.

It appears to Professor Pickering that the iron and stony meteorites differ from one another in other ways besides their composition. That some of the former are associated with falling stars, and therefore with comets, certainly seems plausible. That the latter are not associated with them seems probable, and if so, whence can they have come if not from our own Earth?

CHAPTER XXI

THE ancients assumed that the celestial sphere of the heavens was a reality and even considered it a solid sphere of crystal, on which they observed and marked the positions of the various stars, just as to-day the astronomer assumes for the same purpose an ideal celestial sphere of which the observer is the center and in which declination corresponds with latitude on the Earth and right ascension with longitude. They appreciated that altho the stars moved across the sky and apparently with different rates of motion, yet the distances between any two remained unchanged. For this reason they imagined that the stars were fixed on the celestial sphere. The motion of some of the stars about a center or pole which coincided with the Pole Star was observed, and the two poles in the heavens were early assumed. But the most important observation of the ancients was the relative position of the Sun and the stars. A succession of such observations early showed that the stars were gradually changing their position with respect to the Sun or that the Sun was changing its position with respect to the stars.

Thus the stars obviously vary in position and magnitude, yet in ancient times little was conceived as to their nature or possible origin. There was, of course, the fundamental distinction between the planets or moving bodies and the idea of the fixity of the stars, which was early established and persisted for many centuries, even tho Giordano Bruno (d. 1600) vaguely suggested that the suns of space move. In 1718 Halley had announced a shifting in the sky of Sirius, Aldebaran, Betelgeux and Arcturus since the time of Ptolemy's catalogue, and similar conclusions were reached by various other astronomers. But it was only in 1838 when Bessel with the heliometer was able to detect a motion of the star 61 Cygni that it was clearly and conclusively demonstrated that the stars move through space as well as other bodies in the universe.

The stars were supposed by the ancients to be situated on the celestial sphere at a distance greater, it is true, than the planets; yet as they were observed year after year in essentially the same positions they were held as fixed and immovable and a bodily rotation of the celestial sphere itself was assumed. Just as the ancient mind had given to the planets the names of gods and goddesses, so the wise men of antiquity assigned to the stars similar names or those of animals, the natural result of their vivid imagination. This they did also with groups of stars with even greater play of the imagination.

The idea of grouping the stars into constellations dates from the earliest times. In fact the names long ago given to many have persisted until to-day, tho it must be confessed that they have often proved a cause of embarrassment to the student of astronomy. The modern mind finds it difficult to group in imagination a series of bright points of light in the shape of some mythological hero, bear, dog, serpent or other animal. In most cases the choice had been made in a most arbitrary manner, and Sir John Herschel has truly remarked: "The constellations seem to have been purposely named and delineated to cause as much confusion and inconvenience as possible. Innumerable snakes twine through long and contorted areas of the heavens where no memory can follow them; bears, lions and fishes, large and small, confuse all nomenclature."

The names of the constellations as we know them are

.doubtless of Greek origin, borrowed from Chaldean and Egyptian astronomy. For the most part the names are Greek. The most important are those through which the ancients believed that the Sun passed in its annual circuit of the celestial sphere, or, in other words, those through which the ecliptic passes. For thousands of years these constellations have been used to identify the position of the Sun, especially as the Sun, Moon and five planets were always to be found within a region of the sky extending about 8 degrees on each side of the ecliptic. To this strip of the celestial sphere the term "zodiac" was given, for with one exception all of the constellations it contained were named after living things. It was divided into twelve equal parts, forming the familiar signs of the zodiac, through one of which the Sun passes every month. These signs were made up of a number of stars grouped into constellations. Their names may still be seen, with but unimportant changes, in a modern almanac just as they figured in early Greek days. The names as given in Latin are Aries, Taurus, Gemini, Cancer, Leo, Virgo, Scorpio, Sagittarius, Capricornus, Aquarius and Pisces.

Just how the stars were originally grouped to form the constellations by the ancients history does not record. In an article in the Scientific American it is stated that "the first reliable information regarding the Greek sky is obtained from Eudoxus of Cnidus, an astronomer who lived about 370 B.C. His work furnished Aratus, who lived a hundred years later, with material for his great astronomical poem.

" 'In great numbers,' says Aratus, 'and in various courses the stars incessantly move around the motionless skies. The axle stands immovable. In the midst the Earth is suspended in equilibrium, while the heavens swing around it. The poles bound the axle on both sides. These are encircled by the Bears, that revolve around back to back separated by the Dragon's manifold coils.' "

Eratosthenes (about 170 B.C.) enumerates these constellations, and not only tells the mythological stories but indicates the positions and numbers of stars in every figure, differing from Aratus only in a few particulars.

"Ptolemy gave forty-eight constellations. The figures were the same as the old constellations of Aratus with a few additions. The stars, however, were marked in their proper places and defined as to latitude, longitude and magnitude.

"After Ptolemy a long period ensued during which the astronomical charts were unchanged. It is to the Arabs in the eighth century that the next advance is due. The Caliphs of this period, among whom was Haroun al Raschid of 'Arabian Nights' fame, were friends to science and gathered around them men of learning, such as the famous astronomers Ulug Bekh, Fergani, El-Batan and Abdelrahman Sufi. To a great extent they were satisfied with Ptolemy's work, and altho they retained a great many of the Greek star names, they added a number derived by tradition from the ancient Arab names. Abdelrahman Sufi wrote a detailed and exhaustive account of the Greek constellations, carefully following Ptolemy, and at the same time he treated of the ancient Arabian heavens.

"So strong was their objection to the personal element that when the Greek Zodiac was incorporated by the Arabian astronomers they indicated the names of the objects carried by the characters instead of the characters themselves. Thus Virgo was called the Ears, on account of the wheat she held in her hand; Sagittarius was not the Archer but the Bow, and Aquarius not the Water-bearer but the Well Bucket.

"When the great mixture of Arabian folk-lore was combined with the Greek sky many of the star names were retained, but occasionally the Greek names were changed; for instance, the beautiful red Antares in Scorpio was approximately called the Scorpion's Heart.

"In 1433 Ulug Bekh made at his observatory in Samarkand the most correct catalogue of stars up to that period.

The famous astronomical tables compiled under Alphonso X. of Castile date from 1252, and next in importance was the great catalogue of Tycho Brahé (1546-1601).

"The southern hemisphere, which was uncharted by the ancients, is of far less interest than the northern, partly because the changes have been frequent and unimportant and partly because the only constellation visible to us is the Dove, introduced early in the sixteenth century. In

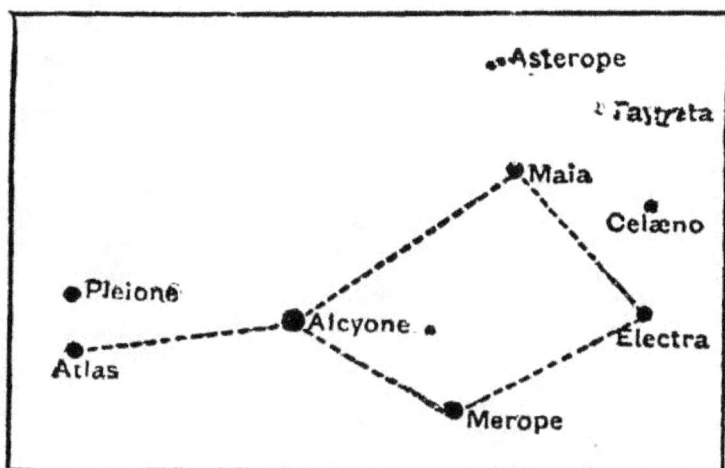

Fig. 38 — THE PLEIADES.

the old books it is called Columba Noæ because it is near the ship, represented sometimes at that period as Noah's Ark. The regions around the Ship and the North Pole have been subject to the most frequent changes since the seventeenth century.

The most familiar constellation is the "Great Bear," known to Americans as "the Dipper"; three stars form the tail of the animal and four others part of his body. To the more prosaic American mind the analogy of the dipper seems far more apt than that of a bear. In Cassiopeia, which is across the pole from the Dipper, the brighter stars form the chair in which a lady is seated. In many cases the position of a figure can be reproduced with a fair degree of certainty, altho it is hard to realize how the names were originally given.

Most of the constellations familiar by observation or legend are those of the northern sky, because until within modern times there were no recorded observations of the southern heavens. Two of the southern constellations, however, are noteworthy, one of which is the Centaur, containing two first-magnitude stars, Alpha and Beta Centauri, the first of which, as we shall see, is notable

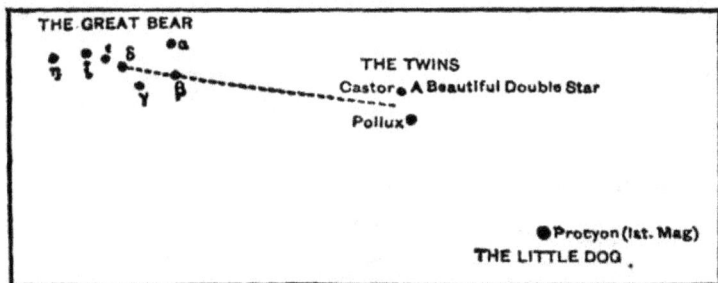

Fig. 39 —CASTON AND POLLUX.

as being the nearest of all the stars to our Earth, while the second constellation is the famous Southern Cross, or Crux, having also a first-magnitude star, Alpha Crucis. Some 5,000 years ago the Southern Cross rose above the English horizon and was just visible in the latitude of London, but through the centuries it has had a southerly motion and has not been seen for many years even in the south of Europe. Perhaps "the chambers of the South" in the Book of Job (ix, 9) may be this constellation, for the Southern Cross must have been a feature in the sky of Palestine when this book was written.

The designation of individual stars probably antedated the idea of constellations; this we may infer from the allusion to the star Arcturus in Job (ix, 9). The two stars Castor and Pollux date from classical antiquity. The names of most individual stars now used are of Arabic origin, which fact accounts for the number of names not Greek or Latin. Thus Aldebaran is a corruption of Al Dabaran, the follower. The modern system of naming stars, however, consists in identifying them with the constellation and then in giving them a separate designation by adding a letter of the Greek alphabet. Thus **the** brightest star of a constellation is called Alpha, the

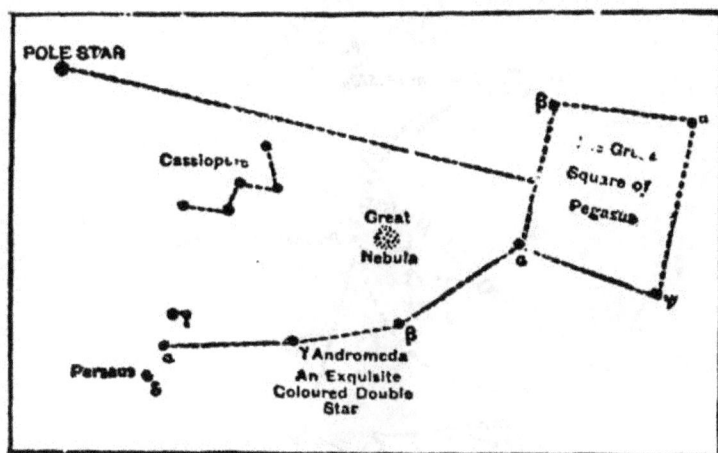

Fig. 40 —THE GREAT SQUARE OF PEGASUS.

next Beta, etc. This rule, which was devised by Beyer for his Uranometria, or star catalogue, published in 1601, has not been followed in all cases. When the number of stars was such as to exhaust the Greek alphabet the Roman was employed and in some cases italics. Flamsteed, the first Astronomer Royal of England, in his catalogue of stars made from observations at Greenwich (1666-1715), introduced a system of numbering the stars. In modern star catalogues both the Bayer letter and the Flamsteed number are often found.

Of the individual stars, tho not the brightest, perhaps

the most important to us is Polaris, the "North Star" or "Pole Star," which is in a straight line with the two stars marking the bottom of the Dipper, which are termed "the pointers." It is five times as far away as the interval between the pointers and very nearly occupies that point of the heavens toward which the north pole of the Earth's axis is directed. To the observer on the Earth it appears

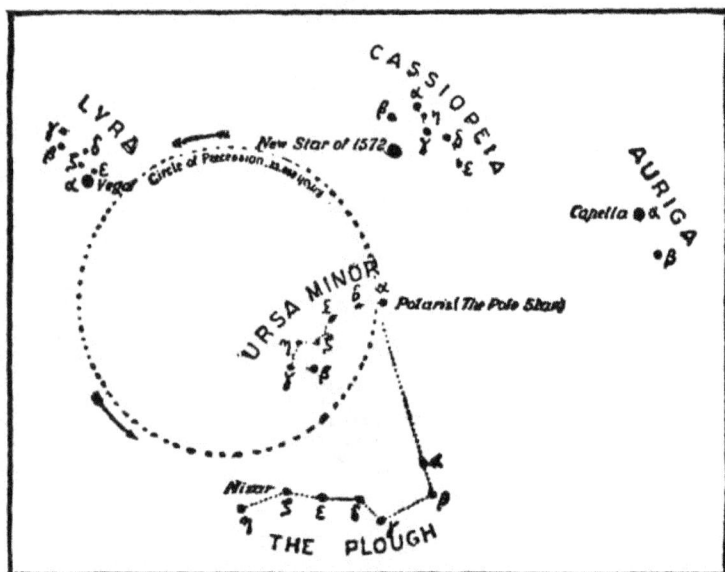

Fig. 41 —POLARIS AND NEIGHBORING CONSTELLATIONS.

to mark the north. Its position, however, varies with its latitude. It has a certain circular motion, due to the slow shifting of the direction of the Earth's axis, known as precession, so that in the course of some twelve thousand years it will be displaced from its position as pole star and Vega, a pale blue star of the first magnitude in the constellation Lyra, will assume its place.

With the conception of the celestial sphere held by the ancients it was not difficult to assume that the stars were

studded about the sky at a considerable distance. Yet
beyond noting their apparent distances from one another
as expressed in angular measure, but little attention was
paid by the early astronomers to their remoteness from
the Earth. Indeed, they assumed them to be very remote,
and Aristotle remarks, quoting the opinion of "the mathematicians," that the stars must be at least nine times as far
off as the Sun. Timocharis and Aristyllus, both of whom
flourished in the early part of the third century B.C., were
the first to ascertain and to record the positions of the
chief stars by means of numerical measures of their distances from fixed positions on the sky, or in other words,
to supply data for the first real star catalogue. But this
did not afford any adequate idea of the absolute or relative
distance.

After the death of Hipparchus it was recorded in an
early text-book of astronomy that the stars need not
necessarily be on the surface of the sphere, but may be at
different distances from the Earth, which, however, could
not be determined with the means at hand. In the works
of this period the conjecture is also hazarded that the Sun
and stars are so far off that the Earth would be a mere
point when seen from the Sun and quite invisible from the
stars. In the Almagest Ptolemy in his series of postulates
(Book I, chap. ii) states that the Earth is merely a point
in comparison with the distance of the fixed stars, and he
believes that it is more rational to assume that bodies like
the stars, which seem to be of the nature of fire, are more
likely to move than the solid Earth.

Copernicus believed that the stars must be at a very
great distance as compared with the size of the Earth,
because the horizon apparently divides the celestial sphere
into two equal parts, and the observer appears to be at the
center of this sphere, no matter how much he moves on
the Earth, so that the distance moved is imperceptible as
compared with the distance of the stars. He maintained
that the stars were all at the same distance from the

Earth, and according to his theory some annual motion of the fixed stars should be observed, due to the alteration of the Earth's position in its orbit. As this was not noticed, Copernicus assumed that the distance of the stars was too great to cause any appreciable motion. The instruments of those days were quite incapable of detecting such a small amount of motion, requiring as it did greater refinements of observation.

The positions of the stars were carefully noted by Tycho Brahe on a great celestial globe 5 feet in diameter, which, in those days, served as a star chart. With a quadrant or quarter circle having a radius of about 19 feet, by which the angles could be read to single minutes, he made a number of observations of the chief star in Cassiopeia, as interest in this constellation was aroused by the appearance in it of a brilliant new star in November, 1572. As he was unable to determine any perceptible parallax, he assumed that the star must certainly be farther off than the Moon. This discovery was important as indicating that changes could occur in that far-distant realm of the fixed stars which hitherto was believed to be constant both in its appearance and constitution. This was corroborated in 1604 by Galileo in the case of the new star in Serpentarius, which he demonstrated was at least more distant than the planets. Galileo in his "Dialogue on the Two Chief Systems of the World" brings out the fact that the angular magnitudes of the fixed stars, which were the most difficult to determine, are in reality almost entirely illusory, and indicates a method known as the double-star or differential method of parallax by which the motion and the distances of two stars at different distances from the Earth can be measured. At the beginning of the eighteenth century, tho thousands of fixed stars had been observed and their positions noted, nothing was known of their distance beyond the fact that it was so very great that they exerted no sensible influence.

It was not until the time of Sir William Herschel that a

successful attempt was made to determine the parallax of a star and to ascertain the distribution of stars in the heavens. Herschel assumed that the distances of the stars depended upon their brightness. In the case of a star barely visible with an 8-inch mirror and one just visible with a 4-foot reflector he assumed that the second was six times as far off. By making a series of measurements he estimated, for example, that the faintest stars visible to the naked eye were about twelve times as remote as such a bright star as Arcturus If Arcturus were removed to 900 times its present distance it would just be visible in Herschel's 20-foot telescope, while more clearly seen in his 40-foot instrument, which he assumed would penetrate about twice as far into space. These observations of Herschel were far more productive of results in other fields than the determination of stellar distances.

It was by Bessel's memorable observations that the angular motion of the star "61 Cygni" was determined in 1838. His calculations showed that the distance of this star was about 400,000 times the distance of the Sun, or $400,000 \times 93,000,000 = 37,200,000,000,000$ miles. This was the first direct solution of a problem which long before the day of Aristotle had puzzled astronomers.

If it is difficult to realize the distances of the planets from the Earth and the Sun in the solar system and the extent of their orbits, what must be said as to the distance of Earth or Sun from the stars? For this all previous scales of measurement were deficient, and a new unit of length, itself of bewildering magnitude, had to be devised in order to realize distinctly the immense intervals of space separating the stars from the Earth. Miles or kilometers were naturally inadequate for such a task. A luminous body emits waves which are propagated through the ether at the rate of 186,300 miles a second, or, in round numbers, six billion miles a year. Accordingly the "light-year" was taken as our unit in discussing the distances of the stars. Thus the light from Bessel's star, 61 Cygni, would take more than six years to reach the Earth. But 61 Cygni is not our nearest star. Subsequent investigation showed that this place belonged to Alpha Centauri, which is four and a third light-years from the Earth and probably ten billions of miles nearer to us than any other member of the sidereal system. Sirius is twice this distance, or eight and a half light-years; Vega is about 30 years, Capella about 32 and Arcturus about 100.

But not for every star are these distances known. They are determined by measuring the parallax wherever possible. To quote Dr. Otto Klotz, "Parallax is the apparent displacement a body suffers by change of place in the observer. With the highly refined instruments and methods, notably the heliometric and spectrographic, definite results are being attained for a comparatively few stars of the myriads that strew the sky and mark a milestone in achievement of practical astronomy. Bold indeed was the undertaking when man first attempted to measure the dimensions of the Earth, but bolder was it when he projected his measuring rod to the Sun. What shall we say when we find him rushing off, measuring rod in hand, at the rate of 186,000 miles a second, and, after 4⅓ years, reaching the nearest star, Alpha Centauri, a bright star of the southern hemisphere? This tremendous distance, quite beyond our grasp when stated in miles, is expressed by saying that its parallax is 0".76. Many of the brightest stars are found to have no sensible parallax, while the majority of those ascertained to be nearest to the Earth are of fifth, sixth or even ninth magnitudes."

If the parallax and distance of a star are determined, it is possible to make some approximation of its mass in the case of binary stars revolving in known orbits. In other words, if we can determine the number of seconds of arc which separate the two members of a binary system and translate this distance into millions of miles and then measure the period of rotation, we can find their combined mass in terms of that of the Sun. For example, the components of the double star Alpha Centauri, to which we have referred, revolve around their common center of gravity at a mean distance of nearly twenty-five times the radius of the Earth's orbit. As they require 8 years for a period of revolution, the attractive force of the two together must be twice that of the Sun. With single stars such a computation cannot be made. Knowing their parallax, however, it is possible to estimate from their size and brilliancy some of the splendid stars in the heavens, such as Canopus, Betelgeux and Rigel, and to realize that they are thousands of times greater than the Sun and suffer in comparison by their infinite distances. For it must always be remembered that the brilliancy of a star depends not only upon its intrinsic brightness but also upon its distance from the Earth.

For five other of the visual binary stars data are available for computing their masses. Thus Sirius, which radiates 32 times as much light as the Sun, is supposed to have a combined mass for its two constituent stars of 3.7 that of the Sun; Procyon, 0.6 that of the Sun; Cassiopeia, 1.8; 70 Ophiuchi, 1.8, and 85 Pegasi, 11.3. The average distance of the stars of a pair of those in the above list from each other is 23 times the distance of the Earth from the Sun, or a little greater than the mean distance of Uranus from the Sun. It is probable, however, that most double stars are farther apart than this. But it is evident from the stars considered that the average mass of a pair is 3.5 times that of the Sun, while the average radiating power is nearly six times as much.

CHAPTER XXII

THE MOTIONS AND BRIGHTNESS OF THE STARS

THE term "fixed star," which has survived from ancient times, we have already found is but relative and that all stars have some motion. If the position of one of these so-called fixed stars is noted by observing the time and the height at which it crosses the south point in the sky, and this observation is compared with accurate records which we have, going back nearly 200 years, it will be found that quite a number of the stars are moving slowly across the sky. This movement is termed *proper motion,* which, as seen from the Earth, is at the best an exceedingly minute quantity. Thus one star, and that the most rapid, has moved about the diameter of the Moon in the last 300 years, or an amount which the telescope is quite incapable of detecting. It may be said in passing, however, that the actual observation of fixed stars from the Earth after an interval of, let us say, 100 years, would show more difference. Yet this is due, not to motions of the stars, but to alterations in the direction of the Earth's axis and other causes which give apparent and common motions.

The maximum proper motion is that of the eighth-magnitude star No. 243 of the fifth hour in the Cordova zone catalogue. The star has an apparent drift of 8".7 annually, which would carry it round a circle of 360° in 149,000 years if it moved uniformly at its present rate. This amount can be appreciated when it is stated that two centuries would be required for a change in position equal to the diameter of the Moon. Arcturus, since the time of Ptolemy, has moved more than a degree and Sirius about half as much, the motions of these two stars having been detected first by Halley in 1718. In fact, the late Professor Newcomb says that if Hipparchus or Ptolemy had made exact determinations of the positions of the stars and to-day should rise from a long sleep, Arcturus would be the only star in which they, or the priests of Babylon for that matter, could detect any change in position relatively to the other stars of the heavens. There is no case in which this quantity is as large as a foot-rule seen at a distance of 50 miles, and for comparatively few stars is this motion certainly appreciable. Notwithstanding the extraordinary degree of precision that has been obtained in recent measures of parallax, for a satisfactory solution of the problem we must probably devise some new method of using the spectroscope or some other instrument for determining proper motions.

The study of the proper motion of stars indicates that, on the whole, the stars are opening out from a point near Vega and closing in to the opposite point. Professor Kapteyn finds that the stars may be divided into two great classes having on the average proper motions quite different in direction and magnitude, and that these two systems are moving through each other, one from a point in the south of Hercules and the other from a point in the Lynx. In the latter class the solar system doubtless belongs, for it seems to be traveling toward that portion of the heavens occupied by Lyra.

The problem of determining the actual speed of travel involves not only a knowledge of the proper motions but also of the distance of the stars. The latter quantity, as has been seen, is obtained by noting the change in position or parallax and is known for about 100 stars. The motion in the line of sight can be measured by spectroscopic methods, employing Doppler's principle. The rate of travel of stars through space has thus been ascertained to be about 10 or 20 miles a second.

How is it known that stars are moving in the line of sight? Millions and millions of miles away a star may be approaching the Earth, and yet through the telescope there is no change of position and no appreciable change in magnitude. But here use is made of an ingenious principle, which takes its name from its discoverer, Christian Doppler (1803-1853), a physicist of Prague, who made an experiment with quite another object in view. Placing some musicians on a railway car and taking his stand on the platform, he noticed that the pitch of the sound was raised as the moving train approached him and was lowered after it had passed and was receding. If this could happen in sound, why not in light, altho the vibrations occur infinitely faster, so that the color of a luminous body (equivalent to pitch in sound) would be affected by its motion? In 1868 Sir William Huggins successfully applied Doppler's principle. If a star is coming toward us or receding there occurs a displacement in the spectra of the manifold light waves varying from the fundamental value of the velocity of light, 186,000 miles a second. When a star approaches, the light waves are crowded together; when it recedes they are drawn apart. Now as the number of the vibrations of the light waves and their length determine the color of light or the position of waves in the spectrum, by microscopically comparing spectrum photographs of the same star and noting the displacement of the Fraunhofer lines it can be determined whether that star is approaching or receding.

Doppler's law has been explained very simply by Prof. Edgar Larkin as follows: "In the diagram (Fig. 42) A is a ray of light from the star S, falling on the side of the prism P, which has the property of separating any mixture of light into separate waves. Light from the Sun or stars is made up of a vast number of colors, all appearing between the limits red and violet. Had the pencil A not encountered the glass, it would have passed to B. But the prism separates the white light into colors which can be projected on any white surface. The red is invariably bent out of its original direction the least, the violet most, and will respectively pass to R and V, with every other color between. The shorter the waves the greater their deflection from a straight course. Red waves run 33,000 to an inch and violet 64,000. An eye at E would see all the colors between R and V direct and at H by deflection, if a screen is allowed to receive the light from R to V. Solar and stellar spectra are crossed at right angles by black Fraunhofer lines. Take any one, say F, anywhere in the spectrum and measure its position with a micrometer. Then the eye, either at E or H, would see a spectrum

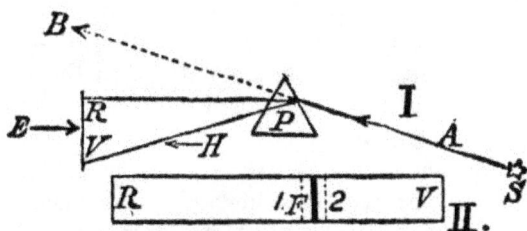

Fig. 42 —DOPPLER'S PRINCIPLE.
The pencil of light, A., coming from the star, S., would pass to B. if the prism, P., were removed. The glass separates the stellar light into all the colors it may contain. The red rays are bent to R. and the violet to V., forming with the colors between, the spectrum of the star.

as outlined in the upper diagram of Fig. 42, extending like R, 1, F, 2, T, V. Let the prism move at great speed, such as that due to the velocity of the Earth, toward the star, or let the star move toward the Earth. Then the line F will move to 2, or toward the violet end. But if the Earth and stars move in opposite directions, the line F will move to 1, or toward the red."

After the more brilliant stars of the heavens had been identified and their positions determined with as much accuracy as possible with the methods of observation and instruments available to the ancient astronomer, the next development was to compare their relative brightness. In 134 B.C., at the time of Hipparchus, a catalogue of stars was prepared which was said to have been suggested by the sudden appearance of a new star in the constellation of the Scorpion. The stars were divided into six magnitudes, according to their brightness. The catalogue of Hipparchus, containing as it did 1,080 stars, has since proved most valuable to astronomers. Next came Ptolemy's great 'Almagest,' published in 138 A.D., which contained a catalogue of 1,028 stars, doubtless based on that of Hipparchus. Ptolemy used a scale of stellar magnitudes, which has continued in use to the present time. The brightest stars of the sky, such as Sirius and Arcturus, were regarded as of the first magnitude, and the faintest visible to the naked eye the sixth. To-day a small telescope will render visible stars down to the ninth magnitude, of which there are over 100,000. Ptolemy employed the first six letters of the Greek alphabet for this purpose and then subdivided the classes he made. If a star seemed bright for its class, he added the Greek letter μ (Mu), standing for $\mu\epsilon\iota\zeta\omega\nu$ (Meizon), large or bright. If the star was faint, he added ϵ (Epsilon), standing for $\epsilon\lambda\alpha\sigma\sigma\omega\nu$ (Elasson), small or faint. These estimates were carefully made. If Ptolemy's original manuscript were at hand his magnitudes would be useful to modern astronomers in determining the secular variation of the brightness of the stars. But the errors in the various copies and transcripts of the Almagest which have come down to modern times are so great that the positions, magnitudes and identifications of about two-thirds of the stars listed are uncertain. Indeed, the oldest manuscript of the Almagest dates only to the ninth century. The Persian astronomer Abd-al-rahman Al-sufi (903-986) reobserved Ptolemy's stars in 964 A.D. and noted cases where he found a difference. This work survived in Arabic, and translated into French by Schjellerup (1874) is available for modern astronomers Uiugh Begh, who flourished about 1450, also published a star catalogue based on Ptolemy, but with careful measures. In 1580 Tycho Brahe published a star catalogue containing the records for 1,005 stars. A supplement carrying this to the South Pole was added by Halley, who went to St. Helena in 1677 for the purpose of making observations of the southern heavens. In 1690 was published a catalogue by Hevel in which several new constellations were added and which was of interest as containing the results of telescopic observations, so that stars invisible to earlier astronomers could be added.

No important additions to the knowledge of the brightness of the stars was made until Sir William Herschel, the greatest of modern astronomers, brought his powerful telescopes to bear on the heavens. He found that when two stars were nearly equal their difference could be estimated very accurately. He adopted a new system for denoting this difference, using points of punctuation—a period denoting equality, a comma a very small interval and a dash a larger interval. From 1796 to 1799 he published in the 'Philosophical Transactions' four catalogues which covered two-thirds of the portion of the sky visible in England. Two other catalogues of his, preserved in manuscript, have not yet been published. It is interesting to know that these observations of Herschel's were reduced at the Harvard College Observatory under the direction of Prof. E. C. Pickering. Herschel's magnitudes for 2,785 stars, observed over a century ago, have an accuracy nearly comparable with the best work of to-day. His work stood unexcelled for nearly half a century, for no astronomer was wise enough to see how much would be gained merely by repeating such observations. Had observations been thus repeated every ten years and extended to the southern stars, many valuable data as to the constancy of the light of stars would have been obtained and our astronomical knowledge greatly increased.

In 1844 Argelander proposed to modify Herschel's

method by using numbers instead of points of punctuation to denote the intermediate brightness between the various magnitudes, a method known by his name. His catalogue, the great Bonn Durchmusterung (1799-1875), contains as many as 324,198 stars visible in the northern hemisphere. After mere judgment with the eye, it was but natural that some more accurate means should be employed, and various photometers, to which we have referred elsewhere, were eventually adopted for the purpose of gaging stellar brightness.

In 1856 Pogson showed that the scale of magnitudes of Ptolemy, which is still in use, could be nearly represented by assuming the unit to be the constant ratio 2.512, which has been adopted as the basis of the standard photometric scale. Thus an increase of four units in number would express the magnitude corresponding with a division of the light by one hundred, and a sixth-magnitude star would have but one-hundredth the brightness of one of the first magnitude.

Photometric observations have been undertaken for many years and are now in progress on a large scale at Harvard University, at Potsdam and at Oxford. Various forms of photometers are employed. Simple photometric work takes into consideration only the total light of a star in so far as it affects the eye. This light may consist of rays of many different wave lengths. In red stars one color predominates and in the blue another. Hence the preferred method is to compare the light of a given wave length (color) in different stars and then to determine the relative intensity of the rays of different wave lengths in different stars, or at least in stars whose spectra are of different types.

The brightness of the stars may also be measured in a simpler but less satisfactory method by determining the total light in a photographic image, a method open to the same objection as eye photometry. In other words, the rays of different colors are combined and affect the photographic plate differently. Consequently blue stars appear brighter than the red. Still photographic photometry is extensively used and various proportions and corrections are employed so that satisfactory results are obtained.

As a result of modern methods of classification the number of stars of the first six magnitudes visible to the naked eye is about 5,000. These are grouped in the following order: First magnitude, 20; second magnitude, 65; third magnitude, 190; fourth magnitude, 425; fifth magnitude, 1,100; sixth magnitude, 320. It has been estimated that over 100,000,000 stars are visible within the range of present visual and photographic instruments.

The first-magnitude stars number only about 20 and on account of their conspicuous brightness serve as landmarks in the study of the heavens. Their names, constellation, magnitude and color are given in the following table:

Star.	Constellation.	Magnitude.	Color.
Sirius	α Canis Majoris	*—1.4	Bluish white
Arcturus	α Bootis	0.0	Orange
Vega	α Lyræ	0.2	Pale blue
Capella	α Aurigæ	0.2	Yellowish
Rigel	α Orionis	0.3	White
Canopus	α Argus	0.4	Bluish
Procyon	α Canis Minoris	0.5	White
Betelgeux	β Orionis	0.9	Ruddy
	α Centauri	1.0	White
Achernar	α Eridani	1.0	White
Altair	α Aquilæ	1.0	Yellowish
Aldebaran	α Tauri	1.0	Red
Antares	α Scorpionis	1.1	Deep red
Pollux	β Geminorum	1.1	Orange

* When a star outshines a star of the first magnitude it is no longer possible to designate its brightness by 1. Hence the numerical expressions 0.2, 0.3 and —1.4 in the foregoing table.

Star.	Constellation.	Magnitude.	Color.
Spica	α Virginis	1.2	White
	β Centauri	1.2	White
	α Crucis	1.3	Bluish white
Fomalhaut	α Piscis Australis	1.3	Ruddy
Regulus	α Leonis	1.4	White
Deneb	α Cygni	1.4	White

In connection with stellar photometry, it has probably occurred to the reader that photographic charts would prove very serviceable, as the brightness of the photographed image could be used. Such indeed is the case. Astronomers interested in stellar photometry have devoted no little attention to the study of these charts and plates.

CHAPTER XXIII

AFTER the various stars in the heavens were classified according to their brightness or magnitude it became apparent that there were striking and important variations in their brilliancy. When Hipparchus compared his lists and catalogues of the brightest stars in the sky with those of earlier observers, he was duly convinced of the occurrence of changes in their position and brightness. This was strikingly emphasized during his own lifetime, when, as we have seen, a bright star flashed up in the constellation of the Scorpion and then slowly faded away. Such changes as this in part induced Hipparchus and other ancient astronomers to make their star catalogues, for it was realized by them that there were from time to time new stars and a large number of variable stars which, while but a small part of the host of stellar bodies, nevertheless existed in considerable number. Unlike other astronomical phenomena, these variations in stars cannot always be predicted. In many instances they obey no rules, and especially in the case of new stars, or "novæ," they blaze up in sudden glory, remaining bright and then perchance fading swiftly away in darkness. For convenience these variable stars have been classified into groups, all containing prominent examples and differing considerably from one another. These classes may be summarized as follows: (1) New stars, or "novæ," consisting of a few stars that appear suddenly like the star in the constellation Scorpion discovered by Hipparchus; (2) variable stars of long period, fluctuating in light by large amounts during periods of several months; (3) stars whose variations are small and irregular; (4) variable stars of short period; and (5) the so-called Algol variables, which are usually of full brightness and at regular intervals grow faint owing to the interposition of a dark companion between the star and the Earth.

Variable stars have long been known, but only about 250 were recognised by astronomers until photography and spectroscopy were applied to their discovery. Three remarkable discoveries, Prof. W. H. Pickering, of Harvard College Observatory, states, are responsible for greatly increasing the number. "The first was by Mrs. Fleming at Harvard College Observatory, who, in studying the photographs of the Henry Draper memorial, found that the stars of the third type, in which the hydrogen lines are bright, are variables of long period. From this property she has discovered 128 new variables and has also shown how they may be classified from their spectra. The differences between the first, second and third types of spectra are not so great as those between the spectra of different variables of long period. The second discovery is that of Professor Bailey, who found that certain globular clusters contain large numbers of variable stars of short period. He has discovered 509 new variables, 396 of them in four clusters. The third discovery, made by Professor Wolf, of Heidelberg, that variables occur in large nebulæ, has led to his disclosure of 65 variables. By similar work Miss Leavitt, of Harvard, has found 295 new variable. The total number of variable stars discovered by photography during the last fifteen years is probably five times the entire number found visually up to the present time. Hundreds of thousands of photometric measures will be required to determine the light-curves, periods and laws regulating the changes these objects undergo."

The discovery of novæ, or new, temporary stars (the first class mentioned), continued from the time of Hipparchus to the invention of the telescope Four new, or temporary, stars were discovered in the interval between the catalogue of this ancient astronomer and the beginning of the sixteenth century. The Star of Bethlehem may have been of this character. In November, 1572, a brilliant new star appeared suddenly in the constellation of Cassiopeia, which was observed by Tycho Brahe during the sixteen months of its life. During this time it rivaled Venus at its brightest and revived Tycho's interest in astronomy, which at the time was beginning to wane. He wrote at considerable length a description of the star, published in 157, and Kepler subsequently remarked that "if that star did nothing else, at least it announced and produced a great astronomer." In modern times the most striking nova was a star which was discovered in Perseus in 1901 by the Rev. Dr. Anderson, of Edinburgh, an amateur, who in 1892 had discovered a nova in the constellation Auriga. In the sudden appearance of Nova Persei, suggests Arrhenius, we evidently witnessed the magnificent termination by collision of the independent existence of two heavenly bodies.

Typical of the second class of variable stars which exhibit marked irregularities in period and in brightness similar to those of the new stars is Eta in Argus, which in 1677 was classed as a star of the fourth magnitude and in 1687 and 1751 of the second and in 1827 of the first magnitude. Then Herschel found that it fluctuated between the first and second magnitudes. In 1837 it increased rapidly in brilliancy and in 1838 so far outshone the typical first-magnitude stars that its magnitude was denoted by 0.2. But in the following year it declined to the first magnitude and there remained until 1843, when it rapidly brightened until it outshone every star except Sirius (magnitude — 1.7). Thereafter it slowly declined to the sixth magnitude and since 1869 has fluctuated between the sixth and seventh. These observed changes point to a great collision in 1743 and a smaller one in 1838. The smaller collision may be compared to the fall of the Earth into the Sun, which would develop heat sufficient to maintain solar radiation during 100 years. From the older observations it appears probable that the star suffered at least one earlier collision.

Another star of this type is Mira, or Omicron Ceti, which was the first star to be recognised as a variable, having been discovered August 13, 1596, by David Fabricius and described minutely by a Dutch astronomer, Phocylides Holwarda (1618-1651), in 1639. In 1667 its period of about eleven months was fixed by Ismael Boulliau, or Bullialdus (1605-1694), altho it was found that its fluctuations varied considerably. It was described in 1780 by Herschel, who had observed it in 1779, when it was nearly as bright as Aldebaran. Four days later the star was invisible even through his telescope. The maximum brightness of this star varies from the first to the fifth magnitude. At the minimum it falls below the sixth magnitude, becoming invisible to the naked eye, and occasionally below the ninth, so that at its maximum it has 1,000 times the luminosity of its minimum appearance. The spectrum of the star indicates that it is surrounded by three nebulous envelopes. The innermost of the surrounding envelopes is uniformly distributed and the others form a ring with two points of maximum density corresponding with the traces of two eruptive streams. This ring revolves in 22 months and has a linear velocity or rotational period of 14.6 miles per second. Hence it follows that the diameter of the ring is 1.45 times that of the Earth's orbit and that the mass of the central stars is slightly less than that of the Sun. Mira Ceti is typical of most of the variable stars in that they are red and give continuous spectra crossed by dark bands and bright hydrogen lines.

The third class of variable stars comprise those of irregular period, which differ from the so-called "new stars"

in that they recur at more or less regular intervals of a
number of years. These are, for the most part, red stars,
altho there are others that fade away and are even lost
from telescopic vision, tho once seen with the unaided
eye. In some cases they have been associated with faint
nebulosities.

Typical of the fourth class of variables is the star Beta
Lyræ. Stars of this class have a short period measured by
hours and days, and their variability is considered due to
eclipses by darker companions, tho both are self-luminous.
As they have a white or yellow color, dust rings in their
neighborhood are believed to play an important part in
their phenomena, tho less so than in the rays of the red
star Mira.

As typical of the fifth class, composed of variables that
change with almost absolute regularity, Algol, or Beta
Persei, may be cited. This star's variability was first noted
by Geminiano Montanari (1632-1687) in 1669, but it was
more than a century later (in 1783) when John Goodricke
(1764-1786), a deaf-and-dumb astronomer, detected the
regularity of these changes and fixed their period at
very nearly 2 d. 20 h. 49 m. Algol at its minimum
luminosity gives about one-quarter as much light as
when brightest, and the change from the first state to the
second is effected in about ten hours. The Algol type
of stars, including about 25 variables, are white in
color and are characterized by a short period, which in
most cases is less than five days. The change in intensity
of light was first accounted for by Prof. E. C. Pickering as due to a second or dark star which travels about
its primary, eclipsing it at various times. As the dark
star begins to eclipse the brighter, the light diminishes
until the time of greatest obscuration is reached, after
which the normal value is attained. Pickering's theory for
Algol, which normally is a star of the second magnitude,
was demonstrated to be true by Vogel at the Potsdam Ob-

servatory in 1889. Therefore it is two stars, or a "spectroscopic binary."

Galileo and subsequent observers noticed that in many instances stars which appeared to the naked eye as single really were double. When these stars were separated under a high magnifying power it was found that they varied in distance, magnitude and color. Thus, if two stars are almost in the same line of sight, they will appear to the observer to be very closely related, altho one of the pair may be much nearer to him than the other and their proximity merely accidental. Such a pair of stars is known as a "double star," or as an "optical double," and is to be distinguished from a pair of stars at approximately the same distance from the Earth, but so affiliated that they revolve about a common center of gravity. In other words, their relative positions would resemble those of the Moon and the Earth. When thus paired the combination is termed a binary system. The discovery of binary stars was first made by Sir William Herschel. He discovered that there were changes in the relative positions of the two stars which obviously were not connected with the motion of the Earth, but indicated an actual circling movement of the bodies themselves under a mutual attraction. He found that there was a regular progressive change in their motion which indicated that one of the stars was slowly describing a regular orbit around the other. Indeed, it seemed that gravitation had its effect beyond the solar system, as the orbit of each of such a pair of stars was found to be an ellipse with the common center of gravity at the focus. That is, the stars were moving in two ellipses which were precisely similar, except that the one described by the smaller star was larger than the other in inverse proportion to the star's mass. While Herschel was the first actually to see such binary stars, yet their existence had been deemed probable by the Rev. John Michell, who lived a short time before the great observer.

Not only are there double stars to the number of 10,000,

but triple and quadruple stars and even multiple stars in a single system. The distances between these stars generally amounts to from 30″ to ¼″, as double stars nearer than the latter figure can be separated by only the most powerful telescopes. But the spectroscope enables stars much closer together to be resolved, and in fact the important class known as spectroscopic binaries previously referred to can be studied and their physical and optical properties determined by elaborate measurements. Thus Polaris, which in a moderate-sized telescope appears as a double star, including one of less than the second and one of less than the ninth magnitude, is really quite complex, for the brighter star revealed by the spectroscope has three stars very close together and all in circulation about one another. Again, in the case of binaries a spectrum with two sets of lines is seen in the spectroscope. With the telescope the image is single, but nothing can explain the double spectrum except the existence of two separate bodies. Accordingly, connected pairs of stars such as these are known as spectroscopic binaries. Often they make up a system of double stars visible in the telescope. Such an example would be Mizar, in the handle of the Plow, which, seen with a small telescope, appears as a double star, one of its components being white and the other greenish. In reality these two stars are situated so distant from each other that from one of them the other would appear merely as an ordinary bright star. But the telescope shows that the brighter of these stars is again a binary system of two huge suns, revolving around each other in a period of about 20 days.

Powerful spectroscopes now are able not only to resolve what appears as one point into two, but to detect motions of the two suns around their gravitation center. This has been simply explained by Prof. Larkin. He states that by the application of Doppler's principle "the times of revolution can be observed, and when the distant suns have a sensible parallax the distance between them can be determined in miles. With time and distance known, velocities in their orbits follow at once. Then with velocity and distance the mass of both can be computed—that is, both suns can be 'weighed' in terms of the mass of our Sun. For it is known from the laws of gravity and motion how much matter is required at any given distance and velocity to set up centrifugal tendency equal and opposite to gravitation.

"The Doppler principle applied to the discovery of

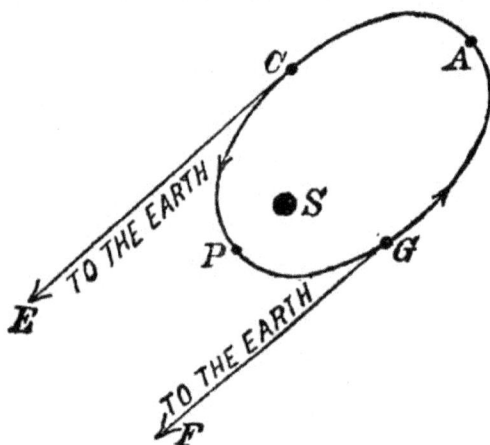

Fig. 43 —A Binary Star.

In this diagram S. is the place of the more massive sun, far and away to one side of the center of the orbit. The orbit and four portions of the revolving sun are represented by A., C., P., G. The two parallel lines, CE. and GF., point toward the earth. When the moving sun is at G., coming toward the earth, a line in the spectrum will shift toward the violet; and when at G., going away from the earth, the line will move toward the red.

binaries is shown in the diagram, where S is a sun, far and away to one side of the center of the orbit of its companion, as shown in four positions, A, C, P, G. As seen from the Earth when at G, the flying sun will send fewer waves per second into the spectroscope and the Fraunhofer lines will shift toward the red. The point A is apastron and P periastron. The velocity of the revolving sun at P in its orbit is enormous, while the light and heat of S are intolerable to the people on any planet revolving around the sun P. For if either sun has a retinue of worlds like the Earth, with inhabitants, their changes of climate are extreme. While the flying sun is moving from A through C and around to P, the huge sun S rapidly increases its apparent diameter and also its light and heat. After passing C in the direction of the arrows, S must look like a blazing globe in skies of incandescent brass. In the revolution of a planet around either sun it would move in between them; there would be no night; the world and its people would be between two white-hot suns; life would expire. But if by chance a few of the inhabitants survive the passage through P, they would freeze when their sun reached the distant point A." Therefore all binaries having great eccentricity of orbits are utterly worthless for support of life on their planets if they have any. All planets are invisible from space, whence it follows that the inhabitants of stellar spaces have not heard of the Earth.

"Let us try Kepler's third law on a binary.

"Suppose an astrophysicist on this speck of dust, the Earth, far away in the direction of the arrows, say 100 or 200 trillion miles—it makes no difference which, provided light from the stars reaches the Earth in quantity sufficient to form a spectrum whose lines can be measured—wishes to find how much matter the two suns in the diagram contain. He sets the telespectroscope on the pair, night after night, and measures with extreme accuracy the shifting of the lines, now toward the red and then the violet. He does not see the stars in the instrument, but their spectra only—tiny delicate and beautiful bands made up of a few colors and black bands. But he watches the lines with great care when the moving sun is at C and G and measures their displacement with the micrometer. By repeated observations he finally learns a thing of vast import, the speed of the Sun's revolution in its orbit, and this entirely by means of the known relation between shifting of spectral lines and the speeds of the flying sun at times of approach and recession. And by direct observation he notes the time of one revolution. He multiplies velocities per second by the number of seconds and thus finds the circumference and radius of the orbit. He at once knows how many times farther apart the two suns are than our Sun and Earth. Then it becomes simple arithmetic to apply Kepler's law and find mass."

The distribution of the stars in the heavens seemed to the ancients to be fairly even on the whole. It was noted, however, that there was a long, relatively thin segment of space extending across the sky in which the stars appeared more numerous than elsewhere, and on account of its brilliancy this was termed the "Milky Way." Aristotle discusses it among other astronomical phenomena, and Greek astronomers looked upon it as a great circle of the celestial sphere. From those days to the present no satisfactory explanation has been offered for the existence of such a segment with the Earth apparently at its center, nor indeed for any of its characteristic peculiarities of aspect and its relationship to the stellar universe. Galileo, with his telescope, found that portions of the Milky Way consisted of multitudes of faint stars clustered together. It was recognised that the stars in the heavens were more and more closely massed as the Milky Way was approached.

With the realization that all stars were not at the same distance, there naturally followed speculation as regards their arrangement and distribution. The first positive contribution to the various theories and the apparent distribution of the stars in space came from Thomas Wright, of Durham (1711-1786), published in his "Theory of the Universe" (1750). This theory is of interest, not so much in its original form as propounded by Wright, but for the fact that it was taken up five years later by the great philosopher, Kant, and by him developed in philosophically explaining the origin of the universe. Furthermore, in the hands of Sir William Herschel it became an important astronomical theory which received serious consideration for many years. Herschel's hypothesis was that the space occupied by the stars resembled in form a thick disk or "grindstone," close to the central part of which the solar system was situated. When such a disk was looked through lengthwise more stars were seen naturally than when it was looked through breadthwise. But at the same time there were vacant spaces or holes in the groundwork of the Milky Way, so that we can apparently see through the collection of stars. These holes or clefts are difficult to explain.

The Milky Way has been described by Prof. George C. Comstock as "a belt approximately following a great circle of the sky, but broad and diffuse throughout one-half of its course, while relatively narrow and well defined on the opposite side. The broad half of the belt is cleft in two by a dark lane running along its axis, and in addition contains numerous rifts and holes, from which the narrow half is relatively free. The number of stars per unit area of the sky is a maximum in the Milky Way and diminishes progressively on either hand, while the inverse relation is true for the nebulæ, their frequency increasing with increasing distance from the Milky Way."

But even the most superficial observer is forced to the conclusion that the stellar system is of limited extent and does not extend to an infinite distance. For if stars and suns exist through an unlimited space, their luminous radiations, which do not suffer in intensity in their passage through the ether, would reach the known universe with little diminution and the entire heavens would always blaze with light. That such is not the case is known from experience and from the fact that the combined illumination furnished by all the stars is only about one-hundredth part of that obtained from the full Moon. The theory has been advanced that the dark holes in the Milky Way, "coal sacks" as they have been termed, consist merely of dark stars or extinguished suns, such as the one we have seen occasions the eclipse of Algol. But this was disputed by the late Professor Newcomb, who says that there is no evidence that the light from the stars in the Milky Way, which are apparently the most distant bodies visible from the Earth, can be intercepted by dark bodies or dark matter, and that the stars are seen just as they are distributed in space. Furthermore, recent photographic work indicates that there is a limit to stellar distribution and that there is "a darkness behind the stars" which long exposure and powerful instruments cannot pierce.

The problem is one of striking immensity. Prof. E. C. Pickering speaks of the distribution of the stars and the constitution of the stellar universe as perhaps the greatest problem in astronomy. In a recent discussion he remarks: "No one can look at the heavens and see such clusters as the Pleiades, Hyades and Coma Berenices without being convinced that the distribution is not due to chance. This view is strengthened by the clusters and doubles seen in even a small telescope. We also see at once that the stars must be of different sizes and that the faint stars are not necessarily the most distant. If the number of stars were infinite and distributed according to the laws of chance throughout infinite and empty space, the background of the sky would be as bright as the surface of the Sun. This is far from being the case. While we can thus draw general conclusions, but little definite information can be obtained without accurate quantitative measures, and this is one of the greatest objects of stellar photometry. If we consider two spheres, with the Sun as common center and one having ten times the radius of the other, the volume of the first will be one thousand times as great as that of the second. It will, therefore, contain a thousand times as many stars. But the most distant stars in the first sphere would be ten times as far off as those in the second sphere, and accordingly if equally bright would appear to have only one-hundredth part of the apparent brightness. Expressed in stellar magnitudes, they would be five magnitudes fainter. In reality the total number of stars of the fifth magnitude and brightness is about 1,500, of the tenth magnitude 373,000, instead of 1,500,000 as we should expect. An absorbing medium in space which would dim the light of the more distant stars is a possible explanation, but this hypothesis does not agree with the actual figures. An examination of the number of adjacent stars shows that it is far in excess of what would be expected if the stars were distributed by chance. Of the three thousand double stars in the "Mensuræ Micrometricæ," the number of stars optically double, or of those which happen to be in line, according to the theory of probabilities, is only about forty. This fact should be recognised in any conclusion regarding the motions of the fixed stars, based upon measures of their position with regard to adjacent bright stars."

CHAPTER XXIV

NEBULÆ AND STAR CLUSTERS

NEBULÆ are masses of diffused shining gas which are scattered through space and which undoubtedly consist of the matter out of which stars have been and are being formed. They differ from star clusters in that the highest powered telescopes yet constructed are unable to resolve them into separate component stars, yet without doubt star clusters are evolved from nebulæ and their connection is most intimate. For the first record of a nebula we go back to Huygens (1629-1695) and find in his "Systema Saturnium" not only a description but a rough drawing. The description is of the nebula Orion and is as follows: "There is one phenomenon among the fixed stars worthy of mention which, so far as I know, has hitherto been noticed by no one, and, indeed, cannot be well observed except with large telescopes. In the sword of Orion are three stars quite close together. In 1656, as I chanced to be viewing the middle one of these with the telescope, instead of a single star twelve showed themselves (a not uncommon circumstance). Three of them almost touched one another, and, with four others, shone through a nebula, so that the space around them seemed far brighter than the rest of the heavens, which was entirely clear, and appeared quite black, the effect being that of an opening in the sky, through which a brighter region was visible."

The work thus inaugurated by Huygens for some reason or other did not seem to attract the attention of astrono-

mers for many years, altho Lacaille, while at the Cape of
Good Hope, 1750-1754, observed and described 42 nebulæ,
nebular stars and star clusters, and altho Charles Messier
(1730-1817), who devoted himself to the detection of
comets, found he was liable to mistake nebulæ for comets,
and recorded in 1781 the positions of 103 of the former. In
the meantime, in 1755 Immanuel Kant, the famous philosopher, advanced the theory on purely theoretical and speculative grounds that a single nebula or star cluster was an
assemblage of stars, comparable in magnitude and structure with the aggregation which we now term the Milky
Way and the other separate stars which can be seen. According to this theory, the Sun would be but one star of a
cluster and every nebula a system of the same order. This
was known as the "island universe" theory and was first
accepted by Sir William Herschel.

In the course of his indefatigable investigation of the
stars with his large telescopes Herschel inaugurated a systematic study of the nebulæ and star clusters. Altho he
found it difficult to draw a line between nebulæ and star
clusters, yet he was able to state positively and correctly
that they were not identical. Herschel noted the position
of each nebula and its general appearance and marked the
positions on a star map. He published catalogues, the first
of which, prepared in 1786, contained 1,000 nebulæ and
clusters, the second in 1789, of about the same extent, and
a third in 1802, comprizing 500. Herschel's observations
of nebulæ enabled him to note their differences in brightness and apparent structure so that he could divide them
into eight classes. In 1786 he published the following
interesting account of the varieties in form which he had
observed:

"I have seen double and treble nebulæ, variously arranged; large ones with small, seeming attendants; narrow
but much extended, lucid nebulæ or bright dashes; some of
the shape of a fan resembling an electric brush, issuing
from a lucid point; others of the cometic shape, with a

seeming nucleus in the center; or like cloudy stars, surrounded with a nebulous atmosphere; a different sort again contain a nebulosity of the milky kind, like that wonderful inexplicable phenomenon about ϴ Orionis; while others shine with a fainter mottled kind of light, which denotes their being resolvable into stars."

Herschel's great problem was to determine the relation between nebulæ and star clusters. Often the difference between the two was made apparent only by the use of a telescope of sufficient power to resolve a bright glow in the heavens into clusters of stars. But at the same time there were bright places that still remained nebulous. Hence Herschel wrote: "Nebulæ can be selected so that an insensible gradation shall take place from a coarse cluster like the Pleiades down to a milky nebulosity like that in Orion, every intermediate step being represented." To Herschel it seemed that the power of the telescope was the important consideration, and the gradation mentioned, he writes, "tends to confirm the hypothesis that all are composed of stars more or less remote."

As Herschel progressed with his investigations the views of other astronomers, as well as those first entertained by him, did not seem tenable. By 1791 he reached the point of view that in certain cases at least the nebulæ were essentially different from star clusters. Referring to a certain nebulous star, he wrote: "Cast your eye on this cloudy star and the result will be no less decisive. . . . Your judgment, I may venture to say, will be that the nebulosity about the star is not of a starry nature." Herschel reasoned that if the phenomenon were due to an aggregation of far-distant stars that there must be one central star of extraordinary dimensions or that something radically different, such as "a shining fluid of a nature totally unknown to us," must be called upon to explain the appearance. His observations proved that an individual nebula was usually surrounded by a region of the sky comparatively free from stars, and that where clusters were

common near the Milky Way nebulæ incapable of resolu-
tion were scarce, but were crowded together in parts of the
sky most remote from this region. In short, Herschel believed that nebulæ and clusters were external "universes,"
and he early believed that both were objects of the same
kind at different stages of development, the result of a
"clustering power" working to convert a diffused nebula
into a brighter and more condensed body, thus indicating
the process of evolution or age.

Berry, in his 'Short History of Astronomy,' to which
we are largely indebted for this record of Herschel's work
in the nebulæ, thus summarizes Herschel's last views of
this important phenomenon:

"His change of opinion in 1791 as to the nature of
nebulæ led to a corresponding modification of his views of
this process of condensation. Of the star already referred
to he remarked that its nebulous envelope 'was more fit to
produce a star by its condensation than to depend upon the
star for its existence.' In 1811 and 1814 he published a
complete theory of a possible process whereby the shining
fluid constituting a diffused nebula might gradually condense—the denser portions of it being centers of attraction
—first into a denser nebula or compressed star cluster, then
into one or more nebulous stars, lastly into a single star or
group of stars. Every supposed stage in this process was
abundantly illustrated from the records of actual nebulæ
and clusters which he had observed.

"In the latter paper he also for the first time recognised
that the clusters in and near the Milky Way really belonged to it and were not independent systems that happened to lie in the same direction as seen by us."

Herschel's observations were utilized by Laplace, who
was engaged in evolving a theory to explain the evolution
of the universe. While his nebular hypothesis in its relation to other theories and systems will be discussed more
fully in the following chapter, yet in this connection it is
desirable to explain how Laplace was able to fit the results

of Herschel's observations to his theory. Laplace had inferred that the planets and their satellites must have been derived from some common source, and he suggested that either they might have been condensed from a body and be regarded as a sun with a vast atmosphere filling the space now occupied by the solar system or that they represented the results of condensation of a fluid mass which now possessed a more or less condensed central nucleus which at one time was not in existence. The nebulæ of Herschel accordingly suggested to Laplace a suitable fluid mass from which a solar system could have been condensed, and, furthermore, the evolution of the fixed stars could be explained on a similar basis. This ingenious theory of Laplace's was rather a scientific speculation than an accurate conclusion founded on data he had himself observed. As a theory, whether accepted or not, it has proved of the most vital importance to science.

John Frederick William Herschel (1792-1871) published a catalogue (1833) of about 2,500 nebulæ, of which some 500 were new and 2,000 were his father's, a few being due to other observers, and later reobserved about 500 known nebulæ while at the Cape of Good Hope (1833-1838), including the nebulæ surrounding the variable star Eta Argus and the wonderful collection of nebulæ, clusters and stars known as the Nebuculæ or Magellanic Clouds. In 1864 Herschel was able to present to the Royal Society a valuable catalogue of all known nebulæ and clusters, amounting to 5,079. Later this great catalogue, which contained a condensed description of each body, was superseded by Dr. Dreyer's general catologue, which was based upon it, and contained 7,840 nebulæ and clusters known up to the end of 1887. A supplementary list subsequently published by the same authority contained 1,529 entries of discoveries made between 1888 and 1894. Hence the two Herschels are responsible for more than half of the total number of nebulæ and star clusters now known to astronomers.

SPIRAL NEBULA IN MESSIER 33 TRIANGULI.

was known as the meteoritic hypothesis and in its more complete form was published in 1890. The fundamental principle involved in this theory was that "all self-luminous bodies in celestial space are composed either of swarms of meteorites or of masses of meteoritic vapor produced by heat." This theory was the result of spectroscopic studies. Lockyer claimed that the original nebula were composed of meteoric material or cosmical dust rather than gases. In other words, nebulæ were vast swarms of meteors and their light resulted from continual collisions between constituent particles. Such a collision might take place between two stars, shattering both and producing a vapor or vapor combined with meteoric fragments from which other stars might be derived. Thus a dark star might be transformed into a bright and glowing star and pass through successive changes until it was dissipated by some long-delayed collision. Lockyer assumed that the meteorites could be regarded as analogous to wandering molecules of gas moving indiscriminately in all directions and at widely different velocities. He believed that the heat produced by the collision already referred to would volatize certain constituents of the meteorites and render them luminous. These luminous materials would supply a spectrum, and the discussion of the properties of the spectra of nebula was an important consideration in Lockyer's work. In some respects this theory has been found untenable. The meteoritic hypothesis was never considered seriously as supplanting the fundamental ideas involved in Laplace's great generalization.

Sir George H. Darwin, whose work on the tides has had considerable bearing on cosmical theory, also discussed the physical and mathematical properties of the meteoritic swarm, and demonstrated that its behavior closely resembled that of a gas and that the meteorites would move about, colliding with one another in much the same fashion as the molecules of gas according to the kinetic theory. Darwin further demonstrated that when a mass was widely extended it would revolve as a solid. But such ideas as those of Lockyer and Darwin involved modifications of the nebular hypothesis rather than an overthrow.

At the very end of the nineteenth century F. R. Moulton and T. C. Chamberlin, of the University of Chicago, denied some of the fundamental ideas involved in the nebular hypothesis. Not only was their work destructive in that they advanced phenomena which the theory of Laplace could not explain, but which were in fact controverted directly by observation, but they constructed a cosmical theory in which these elements were taken into consideration and in which, in large part, the conditions were met. The chief objections to the theory of Laplace, as stated by Professor Moulton, are as follows:

"1. The considerable mutual inclinations of the planes of the planetary orbits and the inclination of the plane of the Sun's equator to the general plane of the system are not to be expected on the basis of the ring theory.

"2. The eccentricities of the planetary orbits are not to be expected on the basis of the ring theory.

"3. The orbits of the planetoids contradict the ring theory.

"4. The rapid revolution of Phobos and of the particles of the inner ring of Saturn cannot be satisfactorily explained.

"5. The presence of light elements in the Earth is not to be expected.

"6. A series of rings could not have been left off.

"7. A ring could not have been condensed into a planet.

"8. The moment of momentum of the present system is less than $\frac{1}{200}$ of that of the supposed initial nebula.

"9. The retrograde revolutions of the ninth satellite of Saturn and (probably) of the seventh satellite of Jupiter flatly contradict the theory."

Of these objections one of the most important is the question of the leaving off of the rings at certain intervals during the contraction of the nebulæ. On this point Professor Moulton writes in his "Introduction to Astronomy": "It is easy to overlook the fact that the postulated nebula must have been excessively rare. According to the hypotheses made, it must have been denser at its center than near its periphery. But if we suppose that it was homogeneous and that it reached out to Neptune's orbit, we find that its density was only $1/_{250000000}$ that of air at the sea level. Neptune's ring could not have been so dense as this, which is many times rarer than the best vacuum yet produced in our laboratories. Now a ring of such rarity would have had no cohesion and would not have separated except particle by particle. When the process was once started it seems that it should have been continuous instead of intermittent as the theory supposes. The theory postulates that when a ring was left behind, the nebula was made stable for a long period of contraction. Roche has attempted to show that rings of considerable dimensions would be abandoned at certain intervals, but his work on this point is far from conclusive. Every other writer on the subject has keenly felt this difficulty. Thus Faye, in his modification of the Laplacian theory, supposed that the whole nebula broke up into rings simultaneously.

"We may assume for the sake of argument," he says, "that rings are abandoned and inquire whether they will unite into planets or not. The matter of the ring would be very widely spread out and mutual gravitation of its parts would be very feeble. The appropriate investigation shows that the tidal forces coming from the interior mass would more strongly tend to scatter the material than its gravitation would to gather it together into a plane. Consequently a ring could not even start to condense into a planet. It would be something like a comet which becomes utterly dissipated by tidal forces.

"To give every possible advantage to the ring theory, we may assume that all the matter has been gathered into a planet except a ring of very small particles, and then ask ourselves whether this minute remainder will be brought to the planet. Investigation shows a strong probability that only that part of the ring which is within 60° of the planet could be brought on to it in any time however long. That is, if we assume that the process of formation of a planet out of a ring is almost finished, we find that it cannot complete itself. This shows the strong improbability that the assumed stage could ever have been reached by condensation from a more uniform ring."

Then it must be considered that the ninth satellite of Saturn and the seventh satellite of Jupiter revolve in a direction contrary to the other revolutions and rotations of the solar system, which would be an impossibility if the Laplacian doctrine were true. Finally, notwithstanding a careful and thoro study of nebulæ, developed by photography to a point of great refinement, there has been found no trace of a nebula which is in the course of breaking up into concentric rings. The spiral nebulæ, such as the theory propounded by Professors Chamberlin and Moulton required, are of the normal type of nebulæ, and it is this fact, brought out in the brilliant discoveries of Professor Keeler, that has largely produced the new theory which has been termed by its originators the "planetesimal theory."

This planetesimal hypothesis to-day is attracting widespread and favorable attention from many astronomers and geologists. While it explains the formation of the solar system and other systems, it does not eliminate the question of primitive matter, which doubtless must have been nebulous. While spiral nebulæ may have been formed by the collision or interaction of large suns, yet these in turn demand for their origin a similar occurrence, so that one is led rather strongly to the belief that the formation of systems of suns and worlds may be a continuous process.

The planetesimal hypothesis has been summarized in the following simple statement by Prof. James F. Kemp in the course of a lecture in which its application to the origin and development of the world is discussed. He writes:

"Instead of a highly heated and subsequently cooled and solidified gaseous original, minute particles of matter, which may have been molecules, are believed to have moved in orbits around a common center in a manner analogous to the solar system of to-day. In their evolution they became aggregated into larger bodies, such as the planets and the Earth, continuing in groups the motions and relations which they possessed when individuals. As the mass gradually increased the pressure of the outer layers consolidated the core and by the mechanical changes involved produced those internal stores of heat with which we are familiar in volcanoes and in deep borings and mines. Vapors or liquids in the original cold particles are believed to have been gradually squeezed out by this pressure. The little particles are called planetesimals or diminutive planets.

"Like all attempts to formulate primeval conditions, the data of this new conception are partly matters of observation, partly assumptions. Speculation enters in a very large degree, and, as in the case of various and widely differing estimates of the age of the Earth based on assumed rates of cooling, once the data were provided, mathematical reasoning goes to a conclusion with unerring accuracy. But the correctness of the solution turns on the reliability of the original data, and where these are so largely assumptive the conclusions are from time to time subject to change. Yet we must have a starting point, and the striking contrasts of the older and later views cannot but impress every one who reflects upon them. The former postulates a highly heated original, the latter a cold one. The one begins with gaseous matter, the other with solid. The one draws upon an original but diminishing store of heat, the other develops heat continuously by mechanical processes. In many ways the two are diametrically opposed; yet some have raised the question whether, in order to obtain a swarm of separate cold particles, we must not in our thought go still farther back to a gaseous or nebulous source, and it is not clear that we have yet escaped the necessity of at least the essential features of the nebular hypothesis."

Whatever theory is studied or adopted, it is clear that the building of suns and the building of worlds is a process of evolution in which the original matter must undergo transformation. The process may be continuous and may extend through infinite time. The collision of suns may have produced nebulæ and these nebulæ in turn may gradually develop themselves into suns again. It seems reasonably certain that nebulæ are the stuff from which the stars are made. Of the many types of nebulæ the spiral forms are considered the more primitive. There is indeed evidence to show that even such apparently irregular nebulæ as that in Orion still preserve traces of original spiral formation.

Professor Dewar has experimentally proved that, at the low temperature prevailing in the outer regions of a nebula, it is one of the properties of a dust particle to attract and condense on its surface whatever gas may be within its immediate vicinity. It may be safely assumed, therefore, that each particle of dust hurled out by the explosion of two colliding suns bears its charge of gas. That these gas-laden dust particles should collide with one another would follow from their motion and their excessive number, and that these particles should be cemented together by the film of liquid gases on their surfaces would in turn follow from the fact of their collision. Thus, in the course of centuries, large grains or aggregations of matter constituting meteorites are formed, which, by gravitational attraction, gather about them part of the remaining dust and gas. Stars, comets and meteorites from other systems will also wander into the nebula, attracting the gas and dust of the nebula. Evidence of this draining of a nebula by an immigrant star is to be found in many a constellation, notably in the Swan, where a distinct lane has been plowed through a nebula by an errant star. By this upbuilding of meteorites from colliding gas-laden dust particles and the concentration of nebular matter by bodies which have strayed into the nebula, the entire mass of gas and dust is converted in the lapse of ages into clusters of stars, slowly revolving about the central Sun in periods of thousands of years. Such star clusters are familiar objects in the heavens. They may be found in the Pleiades, in Pegasus and in other constellations.

Each star in a cluster may be regarded as either the center of a new solar system or as a new-born world—not a rock-bound, sea-swept world as yet, such as the Earth, but a glowing chaotic mass, an embryo which will cool and shrink as it cools and which will eventually chill into a solid sphere. The central Sun, because of its greater size, will blaze on for centuries after the clusters about it have congealed into planets. Because the shrinking and cooling are not of a year's nor even of a century's duration, but are extended over millions of years, it is possible to study the various stages of the process in stars that have been formed from nebulæ at widely different epochs. The method of investigation bears some relation to the system devised by the evolutionists in tracing the development of life on this Earth. With the aid of fossils dug out of the Earth the paleontologist has ingeniously bound the living present to the dead past with the chain of evolution, and shown very convincingly how the creatures that now roam the Earth are inevitably linked with the extinct animals of prehistoric time. An analogous system of stellar classification, formulated by Prof. W. W. Campbell, throws not a little light on the ancestry of the myriad of stars whose rays pierce the heavens. Such a search indubitably points to nebulæ as the stuff of which worlds are fashioned.

In estimating the period which probably elapses before a star in a cluster will shrink into an incrusted world the color of the star's light plays no small part. Red-hot iron is not nearly so hot as white-hot iron. The color of the molten metal gives the iron-founder a clue to its temperature. Similarly a star's temperature may be gaged by its tint. Since temperature, and therefore color, alter with age, it is possible to state in a rough way that red stars are the oldest of luminous bodies and that the orange, yellow and white stars follow in respective antiquity. Youngest and hottest of all are the bluish-white stars.

The spectroscope draws nicer distinctions than those which can be drawn by color alone. Professor Campbell has discovered from his study of stellar spectra that when the spectrum of a star is rich in the bright lines of the gases hydrogen and helium it is certain that the star in question is young. Indeed, the presence of nebulous masses about many stars in this stage of their evolution proves how true is Professor Campbell's conception. In the constellation of Orion, for example, is a nebula in which stars are plunged. The spectra of the nebula and of the stars exhibit the same lines. Can there be any doubt that the stars were formed from the nebula?

In passing from the white stage of infancy the star shrinks more and more. The result is a change in the spectrum. The metals calcium and iron appear in their characteristic lines. Vega and Sirius are stars of this type. As the star ages the hydrogen lines thin out and the lines of the metals become stronger and more numerous. When a stage is reached corresponding with that of our own Sun, a yellow stage, which may be considered the very summit of stellar life, some twenty thousand metallic lines appear. Then comes a gradual cooling. As it changes in hue from yellow to red, more complex chemical combinations are found in the star, and carbon makes its appearance. Then follows a stage represented by the planet Jupiter, still gaseous, still hot, but no longer markedly luminous without the aid of the Sun's borrowed light. A further step brings with it external solidification, the formation of a crust enclosing a hot interior, at which point the Earth has arrived. The last and most pathetic period is represented by the Moon—frozen, desolate, dead.

FRANCIS ROLT-WHEELER

(December 16, 1876 - August 21, 1960)

Francis Rolt-Wheeler was an English writer, astrologer, occultist and esotericist (London 16 December 1876 - Nice 21 August 1960).

After fleeing his home at the age of 12, he embarked for the United States as a sailor on a sailboat. In the United

States, he studied theology at the University of Chicago. He earned a Ph.D. Ordained Anglican priest in 1903, he served for 20 years as a hospital chaplain in New York. In 1923, he began traveling: in Tunisia at Carthage. He founded there the *Institut Scathologique de Carthage*. He settled in Nice-Cimiez (Alpes-Maritimes) and married Yvonne Bélaz. He devotes himself to the writing of works of esotericism and popularization (astrology and occultism). He is the author of the term astrosophy[3] and founds Astrosophy, an esoteric review. The number of his writings in the field of esotericism made him compare to Paracelsus. His wife began to classify the marks left by her husband for future publication.

Works:

- *The science-history of the universe* (1909) Ouvrage collectif, chapitres Anthropology et Ethics

- *The U.S. Service Series*, 20 volumes, Lothrop, Lee & Shepard Company, New York, entre 1909 et 1929.

- *The Museum Series*, neuf volumes, Lothrop, Lee & Shepard Company, New York, entre 1916 et 1927.

- *The Aztec-Hunters*, Boston, Lothrop Lee, 1918.

- *The Wonder of War Series*, quatre volumes, Lothrop, Lee & Shepard Company, New York, entre 1917 et 1919.

- *Round the World with the Boy* Journalists Series, cinq volumes, George H. Doran Company, New York, entre 1921 et 1924.

- *Romance History of America* Series, quatre volumes, George H. Doran Company, New York, entre 1921 et 1925.

- *History of Literature - English Poetry* (The Outline of Knowledge, Volume XI), James A. Richards, Francis Rolt-Wheeler, Edward J. Wheeler, Geoffrey Chaucer, Robyn Hode, Richard Rowlands, Thomas Nashe, Edmund Spenser, Michael Drayton, John Lyle ; J. A. Richards, The Kingsport Press.

- *In the Days Before Columbus*, George H. Doran company, 1921 environ.

- *The Quest of the Western World*, G.H. Doran Co., New York, 1921.

- *The Magic-Makers of Morocco*, Doran, New York, 1924.

- *In The Time of Attila*, Lothrop, Lee & Shepard, New York, 1928.

- *The Pyramid* Builder D. Appleton and Company, 1929.

- *Mystic Gleams from the Holy Grail*, London, Rider & Co4.

- *Summa astrologiae*, en trois volumes. Édition Astrosophie, Nice, 1936.

- *Le Cabbalisme Initiatique en trois volumes*, édition Astrosophie, Nice, 1940.

- *Le Jour de Brahm* en deux tomes, Éditions Adonais, Nice, 1947.

- Le christianisme Ésotérique en deux volumes

- Les douze talismans de pouvoir, édition privée, unique et personnalisée, rédigée uniquement sur commande pour une centaine de souscripteurs entre 1950 et 1960.

FEEDBACK

Now that you have read the book ...

Was it interesting?

Did you enjoy what you wanted to read?
Was there any room for improvement?

Let us know at:
http://www.diamondbooks.ca/feedback

Your feedback is highly appreciated.
Thank you!

Would you like to buy a copy of
'GREAT ASTRONOMERS' ?

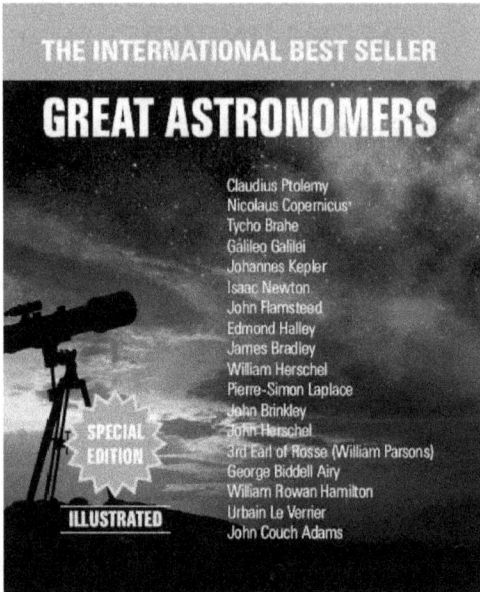

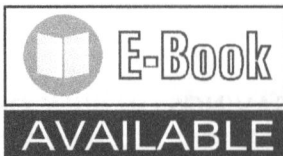

Please visit:
http://www.diamondbooks.ca/books

Would you like to buy a copy of
'GREAT ASTRONOMERS' ?

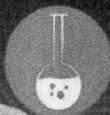

E-Book

AVAILABLE
FOR ALL PUBLISHED TITLES

Please visit:
http://www.diamondbooks.ca/books

DIAMOND
BOOKS
www.diamondbooks.ca

HUGE SAVINGS ON BULK ORDERS
(10 copies, 20 copies, 50 copies, 100 copies, 500 copies, 1000 copies)

Please send your request at:
http://www.diamondbooks.ca/bulkorder

www.ingramcontent.com/pod-product-compliance
Lightning Source LLC
Chambersburg PA
CBHW060321200326
41519CB00011BA/1797